W9-AZU-205

DICK COLLINS'

TIPS TO FLY BY

Also by Richard L. Collins

FLYING SAFELY

FLYING IFR

FLYING THE WEATHER MAP

DICK COLLINS'

TIPS TO FLY BY

RICHARD L. COLLINS

Delacorte Press/Eleanor Friede

Published by
Delacorte Press/Eleanor Friede
1 Dag Hammarskjold Plaza
New York, N.Y. 10017

Copyright © 1980 by Richard L. Collins
All rights reserved. No part of this book may be reproduced or
transmitted in any form or by any means, electronic or mechanical,
including photocopying, recording or by any information storage and
retrieval system, without the written permission of the Publisher, except
where permitted by law.
Manufactured in the United States of America
First printing
Library of Congress Cataloging in Publication Data

Collins, Richard L   1933–
Dick Collins' Tips to fly by.

1. Airplanes—Piloting. I. Title.
II. Title: Tips to fly by.
TL710.C66   629.132′52   80–10833
ISBN 0–440–01733–5

Contents

Foreword

Flying requires a good balance of mental agility and mechanical grace. Pilots who fly well think quickly and operate the machinery with a fine touch. On a day with reasonable flying conditions a passenger hardly feels any sensation of flight after a trip with a good pilot. Everything goes smoothly. When conditions are not quite so reasonable, or when there's a glitch, a talented pilot has the ability to maximize the things going for the flight and minimize the things going against it. Like a cat, he lands on his feet.

Most people who fly could do so as well as the best. They don't, though, and this is a primary reason that the general aviation accident record is much worse than it needs to be.

One reason for sloppy or lackadaisical flying is found in the training process. While it does cover all the facts a pilot needs and more or less forces the rote learning of those facts through the medium of a written test and checkride, the training system often works on an impersonal basis. Many flight instructors do not have a broad range of personal experience in using airplanes to pass along in addition to what is contained in training manuals and courses. Some make the effort to humanize training, and should be commended for it, but it remains that a lot of new pilots know what is true but don't know *why* it is true. They know something about how you are supposed to fly an airplane, but they don't know *why* you are supposed to fly it that way or what happens if you don't.

A new private pilot summed this up when he said he felt superficially trained because of a lack of exposure to valuable experience. What he wanted was an outline of the things he would learn from experience in his first 1,000 hours of flying. This book is based on experience, and it is my hope that the pilot who mentioned the need will find here 1,000 hours' worth of hands-on flying experience.

Discipline is another part of the equation. This is the only place I'll use that word in this book, because my astute colleague Gordon Baxter chides me for using it too much, and I don't want the book to take on the tone of a sermon. I do ask you to remember it, though, because putting yourself in the proper frame of mind is extremely important in flying. To fly well, we have to demand of ourselves nothing short of the best.

Richard L. Collins
August 1979

DICK COLLINS'

TIPS TO FLY BY

1

On the Ground

Virtually everyone who flies also drives a car, and most of us learned to drive before we started flying. Both cars and airplanes are machines used for transportation and recreation, and there is an inevitable tendency to transfer some habits from the car to the airplane—especially the bad habits. In both cases, the problem begins before the start of the trip.

Before driving, a person's prime interest is finding the car keys. With those in hand, most drivers walk straight to the car, open the door, get in, start the engine, and drive away. If there's any "preflight," it likely consists of a glance at the gas gauge. People routinely drive after drinking, after taking medication, and while suffering from any of a wide range of ailments.

Pilots have been known to launch airplanes in much the same manner. Worry not over personal fitness, look the machine over from a distance while walking across the ramp, hop aboard, start up, glance at the gauges, and fly away. There *might* even be a chart under the seat from the last flight.

Such flights usually work out, but they fail often enough to contribute to the long list of needless general aviation accidents. To get off to a good start and progress smoothly, a flight needs a generous dose of planning and thought. It is true that many things often neglected in flight planning are not critical to every flight, but they can be important to an occasional flight. It is also true that wrong decisions are primary factors in over 50 percent of fatal general-aviation accidents. The bad decisions are often made on the ground, and those made in flight are sometimes influenced by something that was or was not done on the ground before takeoff.

Preflight Planning

Preflight planning suggests a pilot poring over charts, E-6B at hand, electronic calculator nearby, and the pilot's operating handbook waiting to provide precise information on the capabilities of the airplane. This is an exaggeration, perhaps, but it may explain why pilots shy away from preflight activity. It can seem complicated, and after you fly for a while, the value of thorough preflight planning becomes less apparent. The winds-aloft forecasts used in those careful calculations are usually inaccurate, making wastepaper out of a flight plan. And more often than not, we fly airplanes well within performance and fuel endurance capabilities, so the limits given in the pilot's operating handbook may seem unimportant.

Considering how general aviation pilots drift away from careful planning after leaving the training process, it's no wonder the accident rate is bad. We become complacent, take too many things for granted, and fly without a complete understanding of the airplane's systems and performance. The air carriers have a far better safety record, and meticulous preflight planning has a lot to do with it.

This chapter, on the things done on the ground, will not be an exhortation always to go through the tedious planning often associated with preflight deliberations. It is common knowledge that few pilots continue to work wind problems on a computer after training, for example, so there would be little point in insisting on this before every flight. It would certainly be *best* to make every useful calculation and computation before each flight, but failing that, a pilot does need a preflight system that insures that the essential checks will be made.

The key to a successful preflight is to consider *everything*. Leave no stone unturned. Make your own personal list and go through it before each flight, much as you run the airplane's checklist. Proper use of any such list comes in recognizing the item that needs special consideration, or work, and giving it the attention it requires.

A very simple list:

Pilot
Fit
Licensed
Current
Motivated

Airplane
Airworthy
Weight
Balance
Fuel
Performance

Flight Plan
Route
Paperwork
Weather

The Pilot

The pilot, the first item on the list, is very important. A lot of pilots kid themselves about their own fitness to fly, and while only a small percentage of accidents are related to a specific physical or mental disorder, a large percentage are influenced by how the pilot feels.

A basic problem is alcohol, with about 6 percent of the fatal accidents in general aviation caused by inebriated pilots. Accidents where the pilot had consumed a small amount of alcohol or was suffering from a hangover probably account for a big slice of the pie. Alcohol in any quantity is a no-go item. Even the FAA's eight-hour bottle-to-throttle rule is inadequate if the bottle was a large one.

Drugs—legal and illegal—are equally debilitating and dangerous. The "don't" list is long. Don't fly if you are taking any medication that affects the way you feel or if for any reason you don't feel up to snuff. The latter is a subjective judgment and might be colored by the relative difficulty of the proposed flight, but remember that an airplane is the wrong place to be when you don't feel well, whether the flight is short or long, or the weather is good or bad.

Some pilots flirt with alcohol, drugs, and flying when not fit because they drive under the same conditions. This is bad enough when driving, but at least a driver can usually pull over to the side when he feels unfit. There is a tremendous difference between flying and driving. When you take off in an airplane, a commitment is made to perform a demanding set of tasks before the airplane can be brought to a safe stop. If, when some miles from an airport, a pilot discovers it was a mistake to fly, the rest of the flight can be a living hell.

Fit to Fly?

Some years ago I set out to satisfy my curiosity once and for all about flying with medication. The flu had wiped me out for a couple of weeks, and the physician I consulted prescribed some pretty potent medicine. I had little confidence in this particular physician even though he was an FAA medical examiner; he actually suggested that I fly myself to Florida to recuperate. His suggestion did get me to the airport, though, where I found a flight instructor to go for a short flight with me.

I hadn't felt too bad driving to the airport, but when

I settled into the pilot's seat of my Skylane, I felt out of place. The technique on takeoff wasn't bad, but once the airplane was airborne I started becoming disoriented—and it was a clear day. I had to work very hard to keep the airplane straight and level, and the farther we moved away from the airport, the worse I felt.

Preoccupation with the way I felt, and with the absolutely basic process of keeping the wings level, was all the old brain was good for. About five minutes after takeoff I was sweating profusely and didn't have the slightest idea of where the airport was. I was wiped out, absolutely, and certainly didn't feel like trying to find the airport and land. I turned the airplane over to the instructor; he flew us back to the airport and landed the airplane. I went home with a lesson learned. If you feel poorly—and even if an FAA medical examiner says you're fit to fly—stay on the ground.

When Flying's a State of Mind

Some people say that flying is therapeutic for them because they can forget a multitude of earthly problems while aloft. This is okay if indeed the other thoughts can be driven out and total concentration given to the flight. But it has been suggested—with some verification—that major events foul up a person's ability to concentrate on such a complex task as flying. Whether good news or bad, any emotional excitement should be considered good reason to carefully preflight your state of mind. New riches are a lot more pleasant than bankruptcy, a birth in the family is a world apart from a death in the family, and a big raise is a lot better than getting fired. But if any event, good or bad, is allowed to dominate your thoughts, your flying suffers. I'll always remember the fellow who went out for a therapeutic flight after receiving his draft notice. On return to the airport, he multiplied his problems by landing wheels up.

Licenses and Currency

It goes without saying that the proper licenses and medical certificates should be in hand and appropriate to the airplane and the flight to be flown. Not quite so plain is that this requirement is a *minimum.* Preflight your real relationship with the airplane and the flight.

I fly my airplane on almost a continuous basis, and if a flight is to be in ol' faithful, I know my machine's capabilities. But if I'm to fly a new-to-me airplane, I always consider any special conditions that might be imposed on the flight. For example, it might be best to eschew IFR or stay away from small airports until familiar with the airplane. It is important to be honest about these things, because a flight that is comfortable in a familiar airplane can be confusing and even hazardous in one that is relatively strange.

I fly a lot of different airplanes in my business and have found many examples of how relatively simple things can add stress to a flight. Automatic pilots are a good example. I learned long ago—after a lot of confused swooping and dipping—never to use an autopilot in IFR conditions until I have thoroughly explored its characteristics in VFR conditions. Even if you are familiar with the airplane, the flight should be VFR or hand-flown IFR until you are familiar with the autopilot. Same goes for the radio system. Most are simple enough once you become used to them, but it can take a while to learn the intricacies. One day a couple of us left an uncontrolled field in a new Mooney loaded with the latest in avionics, and after twenty minutes we had barely figured out how to operate one of the transceivers. Instead of flying along trying to learn how to work the system, we should have worked it out in advance on the ground.

Checkouts in airplanes are often skimpy. The FAA requirement of three takeoffs and landings within the last ninety days in the same category and class (single-engine land, for example) before carrying passengers is far from adequate. If a pilot in command doesn't have full knowledge of an airplane and feel comfortable in the pilot's seat, he should not fly it. There are a lot of important pages in the pilot's operating handbook that should be committed to memory; it is amazing how many pilots ignore them.

Pilots aren't perfect and never will be, but being honest with oneself before flying is surely a key to survival. If there is any doubt about pilot ability or physical or mental well-being, flying should not be done.

Motivation: to Fly Hands-on

What are the best ways to provide yourself with the motivation to fly the airplane properly and safely? One important way is to assume the role of pilot before you take off. You might be highly successful as a doctor, lawyer, or businessman—all demanding much thought—but to be a good aviator you have to be 100 percent aviator when flying. The fact that you make a lot of money or are highly esteemed in your field doesn't really matter in the airplane. Up there, all that matters is flying.

The consequences of botching the flight should be another motivating factor. The airplane at rest doesn't hurt. You are going to set it in motion. The show is in your hands. A little self-reminder in responsibility never hurts: "Richard, old buddy, if you don't do all this correctly the airplane will turn on you. It'll kill you quicker than you can bat an eye."

The Airplane

After an honest self-assessment, it's time to think about the airplane.

The ideal situation is to own your own airplane, be the only person flying it, and keep the airplane in a T-hangar. This assures that as little as possible can happen to the airplane without the pilot's knowledge.

I run through the state of my airplane before every flight. Nobody else flies it, so if there were any squawks at the completion of the last flight, I know about them. When was the oil last changed? Have the plugs been cleaned in the last 100 hours, and how was the last compression check and oil analysis? I *know* that I keep the required VOR, altimeter, and transponder checks up to date, but it's advisable to check them in the airplane's logs on a periodic basis. Same goes for the ELT battery.

If the airplane is just out of maintenance, an extra

careful preflight is in order. A specific item on an airplane I once owned, a Cardinal RG, was the oil cooler duct. It was difficult to place properly when installing the cowling and was worth a special check.

Test hopping an airplane after maintenance can be worthwhile, too. I used to keep my airplane at a small field about six miles from the airport where I had it maintained. The shop had just finished the 100-hour service on the engine, and I had planned to leave the airplane there until early the next morning, when I'd be taking off on a trip. But next morning's weather did not bode well, and I decided to fly the airplane back to its home field while giving it a short test hop.

I took off to the north, flew the six miles to home base, and landed after a straight-in approach. Time aloft was about three minutes. At the hangar, after pushing the airplane back a few feet, I slipped. What did I slip on? Oil. There was a lot of it dripping off the airplane and there was very little of it in the engine—three quarts, to be exact. The other seven had exited through the filter, which had not been properly installed. The airplane would have been completely out of oil in another minute. Needless to say, I had words with the shop and they were quite apologetic. But the inconvenience would have been mine had I taken off on that trip the next morning without test-hopping the airplane.

If the airplane is one you don't fly all the time, such as a rental or club airplane, its maintenance status might not be available for you to check in advance. It might not even be available in the airplane, as the logbooks are often kept elsewhere. In that case, find out when the airplane was last serviced and satisfy yourself that the machine is on a good maintenance program. It does not always follow, but I think it is true that airplanes that are clean and neat tend to be the best maintained airplanes. Conversely, if it looks like junk, it probably is junk.

Some pilots who rent or use club airplanes fly the airplane a few minutes before starting on a long trip, to satisfy themselves that it is ready to go. A good idea.

Weight: the Airplane for the Mission

Does the airplane fit the mission? Weight is an important part of this consideration, because an airplane's weight limitations are not always apparent. A four-place airplane won't necessarily fly with four people and full fuel. In fact, the certification requirement specifies only that at gross weight an airplane must carry a 170-pound person in each seat, full oil, and enough fuel to operate for one hour at maximum continuous power. As an alternative, an airplane is certifiable if it will carry the required crew and full fuel and oil. Most airplanes do better than these requirements, but almost all demand a compromise between fuel and cabin load. Naturally, this compromise must be worked out before flying.

No kidding allowed when adding up weights. And what to do if the weight totals 50 pounds over gross? Leave some fuel out, cutting into the reserve and perhaps prompting an extra stop? Or look the other way while estimating the weight of a steamer trunk at 15 pounds and weigh the kids in as 98-pound weaklings?

There's no question but that a lot of overloading goes on. When there is a problem with overloading, it is usually due to performance problems—specifically in takeoff and climb. On a cool day with a long runway, this wouldn't seem to be a problem. Why worry over 50 pounds?

First, consider what an extra 50 pounds does to the airplane. For example, if the gross weight is 2,800 pounds and the limit load factor is 3.8 times the force of gravity, or 3.8 Gs, that means simply that the structure can really handle 10,640 pounds— its gross weight times 3.8. Load it to 2,850 pounds and the limit load factor becomes 3.73 Gs (10,640 divided by 2,850). Other factors, such as location of the center of gravity and the spanwise distribution of load (meaning the concentration of weight in the cabin as compared with the weight of fuel in the wings), would influence this calculation. But for a "slight" overload, it makes a reasonable illustration. This indicates that there won't be any serious compromise of structural integrity, though all margins will be a bit slimmer than before.

As far as performance goes, one rule of thumb is that performance deterioration will be twice the percentage of a slight overload. In the calculation above, the airplane would be a bit less than 2 percent over gross, so takeoff and climb performance would deteriorate by 4 percent. This would hardly be noticeable.

This may make it sound reasonable to fly with the extra 50 pounds, but it is not. The gross weight is an operating limitation of the airplane, and the regulations (and many insurance policies) very specifically prohibit exceeding the operating limitations of an airplane.

While the effect may seem slight, there are a lot of reasons why overloading is hazardous. Engines can overheat in a climb at higher weights and landing gear components and tires suffer accordingly. The simple fact is that the airplane was thoroughly tested at its gross weight—2,800 pounds in this example—and if it's loaded to a greater weight, the pilot becomes an experimental test pilot. All margins are reduced, and there is *no way* the flight can be conducted as safely as it could at or below gross.

When the Navy launches airplanes off carriers, weight is so critical that it is calculated and written on the side of the airplane with a grease pencil: The man operating the catapult *knows* how much it weighs, as does the pilot. It is not necessary to go to that extreme in general aviation, but a pilot must be certain that the takeoff weight of the airplane is within limits. On flights from marginal airports you might want it below gross weight, and knowledge of precise weight is imperative.

Balance: the CG Is Boss

What about the center of gravity? Some airplanes are prone to problems with aft loading; best learn about this in advance. Exceeding the aft cg limit can be a very serious matter.

Before flying an unfamiliar airplane, I like to settle into a comfortable chair with the pilot's operating handbook and a calculator. In thirty minutes or so it's possible to run through (and double check) every conceivable loading situation. If every example is well within cg limits, as it is with some airplanes, then I make detailed cg calculations for a specific flight only if carrying an unusually heavy object with a reduced fuel load, or flying with people heavier than my present circle of family and friends.

When about to fly an airplane that has a questionable cg situation in any loading, it is best to make the calculations in detail. Remember, the only way to know that the proposed loading is within limits is to check it.

The placard on a baggage compartment can be misleading. If, for example, it reads "120 pounds maximum baggage," that means only that the structural limit of the baggage compartment is 120 pounds. That is a limitation, not an invitation to put 120 pounds of baggage in the area. Nor is it a guarantee, or even a suggestion, that the center of gravity will remain within limits with 120 pounds in the baggage compartment.

Center of gravity is critical because the stability and the longitudinal handling qualities of the airplane change as the center of gravity changes. The farther aft the cg, the less stable the airplane is longitudinally. The control pressure required to raise and lower the nose becomes lighter, and the airplane's resistance to longitudinal displacements caused by turbulence weakens. Even when operating within cg limits, this should be considered before each flight.

I usually fly a four-place machine either alone or with one other person in front. Thus, I usually fly with the cg near the forward limit, where the airplane is as longitudinally stable as it gets. It takes more pressure on the wheel to move the speed a given amount away from the trim speed than when the cg is farther aft. After flying a lot with the cg forward, you will notice that taking off with it aft causes the airplane to feel and behave differently.

One cold winter day I was about to fly away in my Cardinal RG with a full load but couldn't take off until

I cleaned off the frost. This was done with several cans of automotive windshield deicer, and in a few minutes we were ready to go. The takeoff was normal, but the airplane felt uneasy to me after I retracted the gear in the climb. My first thought was frost. Had I really gotten it off? Had something refrozen? Then I realized that I had climbed into warmer air. It was +5 C outside, and frost couldn't have had anything to do with the feel of the airplane. It was just that the loading, which was within limits, was farther aft than I usually fly it.

When the cg is aft of the limit, an airplane can actually become unstable and longitudinally uncontrollable. This can be disastrous, as was related by a pilot who had a problem after loading too much cargo too far aft. The pilot got a clue to his problem as he started to board the aircraft. When he mounted the step, the tricycle-gear airplane settled back on its tail. He still thought it was okay to fly.

The takeoff roll was relatively normal, but when flying speed was reached, the nose of the airplane pitched up sharply. The pilot had the control wheel all the way forward, but the nose would not respond. He reduced power, which caused the nose of the airplane to drop, but because the descent rate was high and the ground near, he applied more power. The nose pitched back up again. He powered back one more time, and when the airplane neared the runway this time, he attempted to cushion the touchdown with power. It still hit hard enough for major damage, and one more pilot learned the hard way not to exceed aft-cg limits. Get it far enough aft, and the airplane is simply not flyable.

Forward cg limits for certification are determined as much by elevator effectiveness on landing as anything else. For a tricycle-gear airplane, a landing must be possible with the nosewheel at a prescribed height off the runway at touchdown. This is less critical than aft cg loadings, but it is still important.

Zero Fuel and Landing Weights

Some airplanes have a "zero fuel" weight. This is confusing to many and begs for an explanation. Simply, it means that all weight in excess of this number must go into fuel. For example, if an airplane has an empty weight of 3,000 pounds, a zero fuel weight of 3,750 pounds, and a gross weight of 4,250 pounds, the highest allowable load in the passenger and baggage compartments is 750 pounds (the difference between empty and zero fuel weights) regardless of how little fuel is carried. All weight above 3,750 pounds must be fuel. This restriction comes because of wing-bending loads. Too much weight in the fuselage results in too much wing bending in turbulence. The weight of fuel in the wings tends to relieve bending loads, and some airplanes (the Twin Comanche for one) actually get a gross weight increase with the addition of tip tanks. The increase is about equal to the weight of the fuel and tip tanks; it has to be utilized for fuel and can't be used to add load in the cabin.

Landing weight is simply that. No fair landing at a higher weight. If gross weight is 4,250 pounds and landing weight 4,050 pounds, 200 pounds of fuel has to be burned after a gross weight takeoff to make the landing legal.

Fuel Is Time, Not Distance

For years, talk of range has been in terms of miles. This is useless, because it is valid only in a calm wind condition. Gas runs through the pipe by the clock, not by the mile, and thinking about fuel had best be done in relation to time. Time comes first; miles delivered depend on true airspeed and wind.

For example, when contemplating a 500-nautical-mile trip in my Cardinal RG, I would look first at the forecast wind. One day it was 12 knots, a little off the nose. That sounded good for a nonstop, because the airplane could fly for six hours and average 130 knots ground-to-ground; it would accept 30 knots on the nose and still make the 500-mile trip with an hour's fuel to spare. Looking at it another way, I always calculate the required groundspeed with regard to required fuel reserve, and that reserve is never less than one hour at normal cruise. I could fly for five hours and still have the reserve, so for this 500-nautical-mile trip the required groundspeed was 100 knots. If the groundspeed dropped below that, it would be a mandate to change plans and make a fuel stop.

In estimating fuel required for start, taxi, takeoff, and climb in piston airplanes I've long used the following rule of thumb: Subtract half as many gallons from the total supply as there are cylinders on the engine (or engines), then divide the remaining fuel by the estimated fuel consumption during the flight. This works well for unturbocharged engines. Compare it with calculations from the pilot's operating handbook before using it on a particular airplane. To estimate an airplane's speed, it works well to subtract 10 knots from true airspeed to

arrive at the anticipated ground-to-ground speed. For
most light airplanes flying below 10,000 feet, this makes
adequate allowance for climb and maneuvering for landing on trips of average length.

About fuel, I annually make a New Year's resolution
always to peer into the tanks of high-wing airplanes before takeoff. Proving the wisdom of this, a friend bloodied the nose of his airplane after taking off on a dry tank
that indicated "full" on the gauge. The gauge system was
bad. It illustrated how the little things can bite if they are
taken for granted.

Performance Guidelines

For most light airplane flights there is no concern
about takeoff and climb performance. We use airports
that are larger than necessary and usually fly at weights
below gross. But there comes a time when takeoff and
climb performance must be carefully evaluated before
flight. Guidelines are useful. I have used the number
3,500 for most light airplanes that I've owned. If an
airport is less than 3,500 feet long, or if the density
altitude is in excess of 3,500 feet, I calculate takeoff and
climb performance before flying. That's very conservative, but if there's any question, best make sure in advance.

When using takeoff and climb data in pilot's operating handbooks to develop guidelines, remember that the
figures given in the book are superlatives based on a
level, dry, hard surface runway. If everything goes exactly as when the figures were developed, you'd really be

50 feet high over the 50-foot obstacle—with absolutely no margin for safety. In real life, we fly airplanes that have some wear on the engine and that might have a brake dragging slightly, and we fly off on wet grass or uphill runways with tires that might be underinflated. Our technique is not always perfect either, and we need a margin of safety.

When my 3,500-foot guideline sends me to the book to calculate performance, I consider the absolute minimum runway length as 2.2 times the ground roll distance shown for takeoff performance in the pilot's operating handbook. To get an equally conservative figure for up and over the 50-foot obstacle, I double the book figure. This is not as restrictive as it sounds. It means only that on a +30 C day at sea level and gross weight, the Cardinal RG requires a runway that is 2,189 feet long, and the 50-foot obstacle should be 3,530 feet from the starting point. Or, if I want to fly a Mooney 201 from a 2,000-foot long sea level strip on a +30 C day, the takeoff weight must be limited to 5 percent below gross. I couldn't accept less margin.

While working with takeoff numbers, learn the required airspeeds for liftoff and initial climb. These vary with weight (they are lower at lighter weights), and many of the newer pilot's operating handbooks list the speeds for several weights. To get best performance, use the appropriate speed.

This discussion of speeds and required runway lengths for takeoff is appropriate to single-engine airplanes. Takeoff planning in multiengine airplanes is much more complex and will be covered in Chapter 6.

Adequate runway length for landing at the destination airport should be calculated during the preflight.

This requires information from the pilot's operating handbook plus many grains of salt. My Cardinal RG book showed a landing ground roll of 770 feet at gross weight after a 63-knot approach in +30 C air. This would be under ideal conditions, which always seem elusive. As a rule of thumb, I triple that ground roll figure for a minimum runway length. When you calculate required landing length, remember the approach speed on which it was based. If the approach is made at a higher speed, the number is no longer valid.

One other item for preflight calculation is the power setting for cruise at the planned altitude and expected temperature aloft. It can be done in advance and it should be done in advance.

The Flight

A complete survey of the route, distances, terrain, and options along the way is very helpful. When flying IFR we generally fly down the airways; when VFR it's best to avoid airways as much as possible. With the rapid proliferation of terminal control areas, planning a VFR route around these and becoming familiar with frequencies in advance is easy enough, while trying to figure out airspace restrictions and TCA frequencies in a bumpy cockpit is difficult.

At Dallas/Fort Worth one day, the clearance out of the TCA was not as I had requested. My idea was to take off and climb to the northeast up to the desired 7,500-foot cruising level. The actual clearance was to fly clear of the lower ring of the TCA at a low altitude. That was

okay, but instead of going back to the chart to determine
any special conditions, I took off and headed northeast.
When the controller said I was clear of the TCA, I realized that I was low, rather close to Addison Airport, and
was about to enter the Addison Airport traffic area without clearance from the tower. I didn't have the foggiest
notion of the tower frequency, and the result was a little
scuffle with the chart to get the appropriate number. It
was a bad lapse in flight planning. I should have reviewed
the change before takeoff.

It takes only a moment to go over things before
flight, to set the stage. I even like to do this for frequently
used routes. An IFR trip from Trenton, New Jersey, to
Asheville, North Carolina, is a good example. There are
standard instrument departures out of Trenton, and one
of these is used in planning the flight because the clearance will surely specify an SID. Trenton Two was the
correct SID for this flight. It took me to the southeast, to
intercept a radial of the Millville Vortac before turning
to the southwest. Then down over Patuxent (southeast
of Washington), where there's a Navy base and a prohibited area that extends up to 5,000 feet. The countryside
is flat, and I had always wondered if air traffic control
would or could approve an altitude lower than 5,000 if
it was needed because of ice or turbulence on the airway
over Patuxent. I had never asked and made it a point to
find out. (Traffic allowing, they would.)

After Patuxent, the airway goes over Richmond,
South Boston, Barrett's Mountain, and into Asheville.
Minimum en route altitudes are 3,500 or lower, and the
countryside remains reasonably flat until about 50 miles
northeast of Asheville. There the 6,000-foot minimum
en route altitude means precisely what it says. The ap-

proach at Asheville usually involves a last-minute rapid descent because of high terrain in the area.

ILS airports along the way are always good alternatives and should be kept in mind. If the weather is good enough to get down anywhere, it's good enough to get down where there's an ILS. Baltimore was off to the right and the airway tracked directly over Richmond. The next were Greensboro and Winston-Salem, to the left. Hickory, right on course, had an ILS, as did Asheville, the destination.

Terrain is an equally important consideration for VFR flights. Every year a lot of VFR pilots press on into adverse weather and fly into rocks obscured by clouds. This is very simple to avoid. All that's necessary is a study of the terrain and obstructions over the course line and a determination of an altitude that will clear the highest point or obstacle within five miles of the flight path by 1,000 feet (2,000 feet in mountainous areas). Write down this altitude at various places on the chart, appropriate to variations in terrain and obstructions along the way. If the flight can't be conducted along that path at that altitude in good VFR conditions, the flight cannot be safely completed. It's that simple. If the flight is VFR but along airways, and if you have IFR charts, the minimum IFR en route altitude should also be considered the minimum en route VFR altitude.

Alternatives should be considered in advance. For VFR, consider all airports along the way and the direction in which the terrain and obstructions will allow safe flight at a lower altitude. For IFR, consider all possible alternate airports.

When it comes to flight log forms, pilots have varying ideas on what is best. My choice is a rather simple one that includes the stations over which I'll be flying and the distances between these stations. If a trip might stretch endurance limits, I always include miles-to-go at each point.

If a detailed log that includes all the airway bearings is what you like, fine. Or if you are the back-of-an-envelope type, that's fine, too. The main purpose of paperwork is to keep up with timing—how long you have flown, how fast you are going, how much longer you can fly—and the primary requirement is to record time off and the times over each point along the way.

Recently a friend of mine got lost because of poor planning, and his little misadventure can be related to flying even though he was driving a car to catch an airliner. My friend had forgotten that his car was an Avis, not a Hertz, and force of habit put him in the lane to the Hertz check-in. When he got there, the Hertz folks naturally wouldn't take the Avis car. He had to drive back out, come around, and make another approach to the terminal. One wrong turn was made before getting to the correct lane, and the upshot of the whole thing was his missing the flight—all because he didn't consider in advance what he was doing. An inconvenience for him in a car, but a very real hazard when flying.

Weather—the Nitty-Gritty

The relationship of weather to flying is the subject of an earlier book of mine, *Flying the Weather Map.* Here it is enough to say only that a very careful and thorough consideration of weather is a critical part of the preflight thought process. Make yourself aware of what's going on and write the information down for study, both before and during the flight. And if there is any question about the suitability of the weather for flying, do the most conservative thing that comes to mind.

Once More, for Safe Flying

Look at the preflight check list again:

>*Pilot*
>Fit
>Licensed
>Current
>Motivated
>
>*Airplane*
>Airworthy
>Weight
>Balance
>Fuel
>Performance
>
>*Flight Plan*
>Route
>Paperwork
>Weather

On short flights in familiar areas and airplanes, some pilots might dismiss the entire list without the stroke of a pencil. It can be done with "okay" responses to each item. In most cases several items will be questioned and action will be required.

At the Airport

Preflighting the airplane is often neglected or glossed over. The pilot's operating handbook prescribes a detailed inspection, though, and this should be followed. It's pretty common not to preflight using the handbook's checklist, but it's absolutely necessary to do it in a methodical way. Work with an open mind, too. One instructor told me of a student who carefully looked into the tanks before flight but failed to see that they were empty. The student replaced the caps and said the fuel was okay, and the instructor didn't discover otherwise until they were taxiing out and he noticed that the fuel gauges were at zero.

It is also very simple to skip something when distracted. If, for example, just before you get to the part about draining the sumps, someone opens a suitcase on the ramp and the underwear starts blowing away, you might forget to come back to the sumps. Water in fuel is a major cause of engine failure, so this is something to check and double check. Drain, making sure there's no water. If some is there, continue draining until all water is out of the system. Be especially suspicious if the airplane has been tied outside in the rain or if it was refueled while rain was falling. It's too late to get rid of the water when the engine starts spluttering after takeoff.

The sumps should also be drained after an en route refueling, and each time—before the first flight or after an en route refueling—the stuff drained should be examined for color and odor, even if there is no water. Odor is important, as the kerosene they use for jets smells different from avgas.

Getting kerosene in the tanks of a piston engine airplane is a very real threat. One of general aviation's
foremost pilots once took on a load of kerosene in his
piston engine twin, failed to note it, and wound up in a
ditch off the end of the runway following an unsuccessful
takeoff. In this case he may well have been given the
wrong fuel because his airplane was turbocharged, and
that was placarded on the cowling. To some line personnel, anything marked *turbo* suggests *jet.* If your airplane
has those words painted on, be especially wary. Or remove the words. This confusion has caused enough accidents and incidents to prompt manufacturers to stop
painting the word "turbo" on most turbocharged airplanes.

In another example, line personnel took an almost
empty 100-octane truck to the fuel farm and filled it with
jet fuel. When a couple of airplanes failed to complete
their takeoff successfully, someone sniffed the contents
of the truck and discovered the error.

You might feel foolish sniffing fuel in front of passengers, but assure them that it's for their own good and you
aren't getting a bargain basement high from aromatic
fuel. The odor of kerosene is easy to detect.

The color is important, too. If it's clear and smells
like kerosene it is probably jet fuel. Another thing that
produces a near-clear liquid is a mixture of 80-octane
and 100-octane. If yours is an 80 burner and the fuel
looks clear and smells like avgas, perhaps it is okay, but
be very suspicious and make sure before takeoff. If the
airplane calls for 100, only blue or green is acceptable.

Almost nobody ties baggage down as they should,
and when I'm tempted to leave heavy stuff loose in the

baggage compartment, I recall an incident involving a friend and his Bonanza. He has always carried a rather large tool chest in the airplane, to make minor adjustments if necessary. One day he flew into some severe turbulence: the tool box wasn't secured, and it broke one of the aft cabin windows. Needless to say, he had to land quickly. The trip couldn't continue until the damage was repaired, and he learned to secure the baggage compartment items. I learned my own lesson on this score while VFR in a Cherokee Six, circumnavigating a thunderstorm. I got a little close and encountered a short but severe period of turbulence, and a two-suiter moved forward from the aft baggage area and wound up on the floor behind the right front seat. If you don't tie down everything, secure anything that could hurt if it hit you or that could break a window.

If the airplane has been outside in the wintertime, ice or frost must be completely removed, as the pilot's operating handbook insists. If ice and frost are removed by putting the airplane into a warm hangar, be wary of any moisture that might refreeze when the airplane is put back outside. If water should run into control surfaces and refreeze, the surfaces could be enough out of balance to flutter. When deicing in a hangar, I always spray deicing fluid generously on the airplane before rolling it out, to prevent refreezing.

The pitfalls of not deicing are illustrated by an air carrier mishap in a DC-9. The airplane had collected some ice on an approach, and more formed as freezing rain fell while the airplane was being loaded for the next leg. The captain declined to have the airplane deiced on the ground.

The airplane's acceleration on takeoff was okay, and

it flew off and into a positive rate of climb after the pilot rotated at the proper speed. But as the gear came up, the airplane rolled rapidly to the right. Then it rolled to the left and the left wing struck the runway. Then it settled to the ground and slid off the end of the runway and into the woods. It had simply stalled. The roughness of the ice on the flying surfaces completely fouled up the aerodynamics of the moment. If a DC-9 doesn't have enough power to overcome a little coating of ice, a light airplane has hardly a chance.

Variations for a Tailor-made Preflight

Not only should the preflight check procedure in the pilot's operating handbook be used, it should be improved on where necessary. On my Cardinal RG I added the item about the oil cooler air intake after it was improperly installed a couple of times after maintenance. I also modified the checklist item that read "Stabilator and Rudder Trim—TAKEOFF," and for very good reason.

It was a cold morning. The last time I had flown the airplane was in warmer weather, and the last hour of that flight a couple of days before was in very heavy rain. On this sparkling clear and cold morning, I preflighted, taxied to the end of the runway, checked everything, and was satisfied that the airplane was in good takeoff condition. Once off and with the gear up, the airplane needed a bit of nose-down trim. But I couldn't move the trim wheel. It was frozen solid. No problem, though an airplane in which you have to maintain a constant forward pressure on the wheel to keep the attitude as desired is not a pleasant one to fly. I circled the field, landed, applied some heat to the trim actuator, and went on my way. Water had entered the actuator and frozen.

Now I check trim systems for free and correct.

Once the preflight is completed, it's wise to go through a mental review of the things that are superimportant. If the airplane has uncontaminated fuel and clear fuel vents, it should run. If the caps are properly secured, no fuel should siphon. If it's full of oil, lubrication should be okay. If the prop has no nicks or loose blades, things should be okay there. If the static vents and pitot head are clear, the pressure instruments should work properly.

Checklists Can Save the Day

"Only sissies use checklists."

"If you have to use that list, you must not know what you are doing."

Checklists have been maligned over the years, but their use has withstood the test of time. Even when flying familiar machines, the checklist keeps you on track.

The value of the checklist is apparent when you review the many accidents that have happened over the years because some important item on the list was neglected.

"Controls Free?" A lot of pilots, including airline pilots, have attempted to fly with the control locks on. It seldom works.

"Fuel On?" On some airplanes, if the fuel valve is "off," the engine will operate long enough to get the airplane off the ground. Then there is silence. The fuel selector should be on a tank with fuel.

"Cabin Doors Closed?" More than one airplane has been wrecked because of problems that occurred after an improperly secured cabin door came open.

"Mixture Rich?" In some airplanes, pilots taxi with the mixture leaned to prevent plug fouling. Accidents have occurred while attempting takeoff with the mixture leaned for taxi.

"Flight Instruments Checked?" I know of one accident in which a pilot became disoriented after takeoff because of an erroneous reading on the directional gyro. The pilot subsequently lost control of the airplane while preoccupied with trying to figure out the incorrect DG readings.

These are just a few examples. Anybody who flies airplanes with cowl flaps and doesn't use a checklist might as well admit to forgetting them numerous times. On occasion, we've all forgotten the fuel pumps or other required checks.

When flying a Cessna 310 with a demonstrator pilot a number of years ago, I was made aware of the things you can miss. We had landed and taxied back for another go. I ran a little mental checklist that I use—ICFT, I Can Fly This if the Instruments, Controls, Fuel, and Trim are okay—and glanced at the trailing edge of the wing to insure that the flaps were up.

There was a crosswind, and the airplane became light on its main gear early in the run. Runway handling wasn't good, and the pilot in the right seat remarked that the wind must be stronger than reported. Nice of him to say that. About then, the airplane flew off the runway at a speed not much above the stall. It felt awkward, and I finally realized that the flaps were fully down and started working them gently to the up position. Even though I had given passing thought to flaps, I failed on this point. I had looked at the trailing edge of the wing, but a 310 has split flaps that the pilot can't see when they are down; only the gauge on the panel, or the position of the flap lever, tells for sure.

A heavily loaded jetliner was once lost because of a flap-setting problem. The takeoff began with no flaps, even though use of flaps was specified. When this was discovered, a crewmember set out to rectify the situation but extended full rather than takeoff flaps. There was no way the airplane could fly in that configuration, and the result was a disastrous run off the end of the runway and into the rough.

The simple act of missing one item on the checklist has led to other serious accidents. It's pretty simple: Checklists are not for sissies, they are for smart pilots.

Logical Sequence in Checklists

Some checklists do seem in illogical order. They result in the pilot spending a lot of time running through the checklist either at the number-one position for take-off or at the ramp with the engine running. Items that can be checked before starting, such as the trim, should be checked at that time. If you set out to rearrange the checklist for your airplane to accomplish this, just be sure nothing is omitted from the revised version. Also, any items that are in a logical sequence of necessity should remain in that sequence.

Some pilots are comfortable with a substitute for the checklist, to be used only in familiar airplanes. This takes the form of a right to left sweep across the panel and a vertical check from floor to top, especially up the radio stack. This might be reasonably effective, but I don't think it can be as effective as a checklist.

I have already mentioned a pilot who looked into empty tanks and believed everything was okay because he had gone through the motions of looking into a tank. The same mistake can easily be made with a checklist, or anything else for that matter. For any system to be valuable, it must be used with objectivity. Running a checklist must be a thoughtful exercise, not a mechanical motion. Properly done, it helps eliminate the anxious moments that come from forgotten flaps, frozen Pitot tubes, or overheating engines.

Preflight the Airport and the Surface Winds

The airport and the air around it are also worth a bit of preflight inspection. What about crosswind? If there's a crosswind, spend a moment analyzing how the wind and associated turbulence will affect the airplane during takeoff and initial climb. Close obstructions upwind of the runway can be a factor in takeoff and initial climb, and higher terrain upwind might mean a downdraft at a point when the climb performance is critical. If the wind is 90 degrees across the runway and you are given a choice, it would be much better to climb over flat terrain than over rough terrain.

A 90-degree crosswind is seldom steady at that figure if the wind is gusty. Watch the sock or tetrahedron for a minute before casting a takeoff direction in stone.

What's the crosswind limit of the airplane? Many have a placard listing a maximum demonstrated crosswind component. That number should be considered in context. It is not a guarantee or an invitation, it is just a statement that the airplane was flown in a crosswind of the stated velocity during certification and was demonstrated to be controllable. Crosswind limits are more often set by pilot technique than by the airplane.

If the airport has more than one runway and the wind is across both, the selection of runways might be influenced by things other than wind. If, for example, one runway is narrow and short and has 30 degrees of crosswind and the other is wide and long with 45 degrees of crosswind, the better choice might be the wider and longer runway.

Crossed or otherwise, the wind itself can be a factor. How much wind can the airplane handle? Will taxiing crosswind or downwind present a problem? Again, wind is a matter of technique until it becomes so strong that it overpowers the airplane. Any time the peak gust value is equal to or greater than 50 percent of the flaps-up stalling speed, you can bet that technique plays an increasing role in ground handling of the airplane. In high-wing airplanes, the control positions in the pilot's operating handbook are very important. In any airplane, taxiing slowly and avoiding ice patches (which seem to go with high wind) are essential safety measures.

As far as determining a "wind limit" for an airplane goes, I'll pass. No doubt there is a limit for each and every airplane. This is the wind velocity at which the

airplane will either blow over in a crosswind, weather-
vane into the wind, or just blow away. With the donation
of a few of each type of airplane to the whims of the wind,
these values could no doubt be determined. But it probably wouldn't serve a very useful purpose: Pilots might be
encouraged to operate in wind that was within the limits
of the airplane but beyond that which the pilot's technique could handle. Most FBOs put some limit on the
allowable wind for flight school operations. The rest of
us are left to set our own limits. Certainly if it's more than
half the flaps-up stalling speed we should think twice—
if there is some doubt, there's too much wind.

Gradient Wind

One final item on surface weather and how it can
affect initial climb: When it is calm or there is a light
surface wind, there is often an increase in wind velocity
beginning a few hundred feet above the surface. This can
be seen on a day with low clouds: Look at the clouds and
determine the wind direction from their motion. If you
take off into the wind, as indicated by the movement of
those clouds, the climb rate is likely to get a temporary
boost as the airplane climbs into a rapidly increasing
headwind. Likewise, if you take off away from the wind,
as indicated by the movement of low clouds, the climb
is likely to sag for a bit as the airplane ascends with an
increasing tailwind. If climb is critical, it's quite obvious
which would be the best way to go. This is something to
watch for at any time the surface winds are light and the
forecast winds aloft are strong.

The Airport Environment

If the runway is grass, that should have been taken into account during takeoff calculations. Most pilot's operating handbooks give a factor to add for a dry grass runway. Make sure the grass is dry if that figure is used. If it is wet, add more. If there is standing water, add a lot more.

On paved runways, snow or ice or water can prolong the takeoff run and should also be considered.

After satisfying yourself that the airport is in decent condition, make one more calculation if the runway length is the least bit marginal. Eyeball it for the halfway point. Given that our minimum runway length is 2.2 times the takeoff ground roll as per the pilot's operating handbook, acceleration is below par if the airplane has not accelerated to takeoff speed just before reaching the runway halfway point. In my Cardinal RG, I knew that under most conditions I could accelerate to that halfway point and if not at flying speed there I could abort the takeoff and stop in the runway remaining. The handbook figures for stopping are always about the same as those for the takeoff ground roll. And if I couldn't stop, I would be going very slowly when leaving the runway.

A good example of the nice margins provided by this takeoff technique came one morning when I was leaving a 2,200-foot grass runway which was wet after heavy rain the night before. The airplane was light, and according to the 2.2 rule the required runway length on dry grass would have been 1,960 feet. I drove the runway in a car and determined that, while there were patches of water, I could avoid most of them.

When the halfway point was at my 10 o'clock position, the airplane should have been at its liftoff speed of
56 knots, but it was not. It was just over 50 knots, but it
was accelerating. By not rejecting the takeoff at that
point, I ruled out a completely successful abort. But the
margin I had built into the calculations was generous
enough to make everything work well. If speed had been
down around 40 knots at that halfway point, I'd have
known for sure that I was flirting with real trouble.

Reward

The rewards of competent preflight planning are
realized when you pull into position for takeoff without
nagging questions and doubts about the wisdom of the
flight. Everything has been considered, and everything
looks good. The time required to do the thinking and the
work is not great, and it makes the difference between a
launch based on faith and hope and a launch based on
knowledge and confidence. The latter is much better
when you put your hands on the controls and start to fly.

2

Takeoff and Initial Climb

The takeoff might be simpler than the landing, but first things first. Smooth acceleration to the proper speed, flawless directional control, a perfect rotation to the flying attitude, and an efficient initial climb make the transition to flight graceful and rewarding. The passengers might not be as impressed as with a perfect approach and landing, but there's equal cause for them to be grateful: The number of fatal accidents in takeoff and initial climb is about equal to the number in approach and landing.

Takeoff

Careful treatment of the engine is a big part of flying, and on takeoff this means a smooth application of power. Jamming the throttle home, and asking the engine to go from idle to rated power in a very short time, can result in engine damage or failure. The bigger and more complicated the engine, the greater the need for gentle throttle action. On many turbocharged engines, an abrupt application of power can result in overboosting, which can wipe out the engine. If the runway is short, it's best to hold brakes while applying power smoothly rather than jamming the throttle. Then let it roll.

On short fields, some pilots apply power as they are wheeling into position for takeoff. This acceleration in a turn puts a side load on the landing gear and also causes fuel in the tanks to slosh toward the outside of the turn; if the tanks are not full, it might cause the fuel to move away from the line that feeds the engine, and a power interruption can result.

In preflight, the calculations on takeoff roll assume that the airplane will accelerate normally. It almost always will, too, but the day might come when it doesn't —even on what appears to be a normal takeoff.

An accident involving an airline jet illustrates this clearly. The airplane was loaded to gross weight for a transoceanic flight. The rotation speed for the takeoff was calculated at 153 knots. According to the calculations this speed should have been reached in 5,700 feet and the airplane should have lifted off the runway 6,600 feet from the start of the takeoff roll, at a speed of 163 knots. The runway in use was 10,900 feet long. From calculations everything looked good.

The captain was handling the power and the brakes and making the airspeed callouts; the first officer was assigned to make the takeoff. The captain testified: "The aircraft appeared normal up to approximately 130–135 knots. The speed did not diminish, the acceleration somewhat was decayed or flattened out. I continued to V_1 [138 knots]. V_1 [the speed at which the takeoff would have to be rejected for the airplane to stop in the remaining runway] was reached and there was no more decay, the acceleration was continuing . . . and at 145 knots or . . . somewhere within that area, the speed flattened out, the acceleration flattened out. We continued and it appeared that there was sufficient runway to continue the takeoff, rotate, and continue flight. V_R [the speed at which the aircraft is rotated to a flying attitude] was reached. I called V_R, and this appeared to be approximately . . . eighteen to fifteen hundred feet from the end of the runway. The aircraft was rotated. I followed through [on the controls] . . . and the airplane did not come off."

The first officer stated: "It seemed like it took a few moments longer to get to V_1 than normal. With our rate of acceleration we had and the remaining runway, it appeared to me that there was no problem involved.

"Several times during the run to V_1, I checked the engine instruments. They all seemed to be reading properly, and at the 80-knot call, I checked the engine instruments too, and they were reading normally.

"After V_1 there was a definite lag in the acceleration, but still with the rate it was increasing, it appeared to me there would be plenty of room to reach V_R, rotate, and clear the runway before the end.

"Upon reaching V_R it still appeared to me that we could rotate and become clear of the airport before the end of the runway."

Two passengers, both Air Force pilots, stated that the initial acceleration of the aircraft on the takeoff roll appeared to be slow and that after they had proceeded about 2,000 to 3,000 feet down the runway they began to hear a series of loud reports which they believed were the aircraft's tires blowing out.

The aircraft actually reached a maximum speed of 152 knots on the takeoff, 10,400 feet from the start of the roll. This was 4,816 feet past where that speed should have been reached. The aircraft did not fly, power was cut, and it finally stopped 3,400 feet past the end of the runway after hitting a wooden barrier, the ILS structure, and a drainage ditch.

The problem was that the main landing-gear wheels did not rotate at all on the takeoff. The brakes were locked. The start of the takeoff apparently felt normal because the runway was ice-coated so the airplane had no problem moving. But as speed increased, the tires blew out and the airplane was trying to accelerate while running on the rims. Despite the fact that acceleration seemed okay to both pilots up to V_1, it had taken the aircraft 20.8 seconds and 3,200 feet longer to reach that speed than normal.

This clearly illustrates the value of estimating a point on the runway at which liftoff should occur before beginning the takeoff roll. If the airplane isn't up to speed at that point, you'll *know* that acceleration is not as good as anticipated and that no calculations remain valid. This is

preferable by far to relying on feel for proper acceleration. In the case just related, professional pilots with a
combined total time of over 7,500 hours in the type of
aircraft involved did not perceive substandard acceleration, so most of us should take heed and not rely entirely
on the seat of our pants.

It goes without saying that any sign of engine roughness during the takeoff run should dictate an abort. The
same is true of any unusual noise. It's a lot better to abort
and investigate than to take a sick engine flying, or fly
with something banging around on the airframe.

Directional Control and Takeoff Performance

While the tricycle gear makes directional control on
takeoff much easier than it was back in the taildragger
days, smooth footwork still has its rewards. Steering
effectiveness varies from airplane to airplane, and by the
time you have taxied out to the end of the runway you
should have a feel for how it's going to steer on takeoff.
I prefer the ones that are a little sloppy on the steering.
With steering that's not so direct, you are using the nosewheel *and* rudder to steer, with the rudder doing more
and more of the steering as the airplane accelerates.
When the nosewheel is lifted there's not that much of a
change in steering authority. With direct steering, things
can get a little goosy when the groundspeed gets high on
a takeoff at gross weight or in high density altitude and
calm wind conditions. But to each his own.

Directional control can be influenced by the distribu-

tion of weight on the wheels. This is especially true in a
crosswind (*see* page 48). The primary rule on a normal
takeoff in most airplanes is to let the airplane run naturally—trim properly set for takeoff and the elevator in
the position it seeks. Pushing forward on the elevator
control during takeoff reduces the weight on the rear
wheels and increases the weight on the nosewheel. This
makes the airplane steer like an overloaded wheelbar-
row. Pulling back on the elevator control in the takeoff
roll means that the wing will start creating lift and the
weight on the wheels will progressively lighten. In the
latter case, the aerodynamic drag will be higher in the
takeoff roll and the run will be prolonged. Best to just let
it run as it wants to run, except on a soft-field takeoff
(*see* page 50).

When the airplane reaches the predetermined rotation speed, pull back with the elevator control until the
airplane is in the proper attitude for initial climb. If the
speed at which you begin to rotate the airplane for liftoff
is much lower than initial climb speed, the control movement should be relatively gentle. If the speeds are about
the same, a somewhat brisker rotation is okay.

It's possible to form an advance opinion of what to
expect in the way of control forces at liftoff. If the airplane is one that sits longitudinally level, it's probably
going to take a bit more pull than if the airplane is one
that rests a bit tail-down. If the center of gravity is forward, it's probably going to take more pull than if the cg
is aft. If you can feel the effect of downsprings in the
elevator system, a good pull will probably be required.
The main thing is to have a keen eye for attitude and
move the controls to fly smoothly but surely.

Raising the nose too much, or too soon, can foul up

takeoff performance. This was perplexing to a fellow I taught to fly in a Cessna Skyhawk. He had learned in the wintertime and had always flown the airplane solo or with me in the right front seat. As warm weather came, we took a trip with a couple of other people along. I neglected to brief him on the differences to expect in the takeoff, and he was used to the takeoffs he had been making all winter.

The airplane accelerated more slowly, but he did let it run level to the liftoff speed that he normally used. Then he raised the nose too much. The amount of back pressure he used was probably the same as before, but the center of gravity was farther aft then he was accustomed to, and equal pressure on the wheel resulted in greater attitude change. The nose came up higher than it should have, and the airplane did not fly off the runway as he expected. Instead, it ran along on its back wheels without accelerating. The combination of the weight, heat, and the aerodynamic drag caused by the nose-high attitude was more than the engine could overcome. The cure was to lower the nosewheel back to the runway and allow the airplane to gain speed.

Initial Climb

Once a single-engine airplane is off the ground, an efficient initial climb is at the best angle-of-climb speed, landing gear up where applicable, full power, with the flaps left at takeoff setting. Normally I hold this configuration until the airplane is 1,000 feet above the ground. It affords the most altitude in the least possible forward distance. Also, a position close to the field, from which more alternate landing sites are available, is reached sooner. Altitude is the best noise muffler, and leaving the airport area at maximum means being nice to the neighbors, too.

Another good reason to leave the engine at takeoff power until 1,000 feet is the oft-quoted fact (based on a rather old FAA study) that most engine failures occur when the power is changed. The reason for this is that at a time of power setting change the internal stresses of the engine change, and something that was just about to let go will do so then. The power change doesn't promote the engine failure, but it does affect the timing of it. If we accept this as fact, and there's no reason not to, then it's not smart to do anything that might prompt an engine failure until you are ready to deal with it. If flying an engine that is approved for continuous operation at full power, climb to 2,000 wide open if you wish, or 3,000.

If the preflight was done properly, and if the airplane has been maintained properly, the chances of an engine failure during initial climb, or at any other time, are slim. But what if it does happen? Should you go straight ahead or turn and return to the airport? It depends on a lot of things.

A pilot I know was faced with this one day when his engine quit on climb-out, and his options were bleak indeed. Going straight ahead meant apartment houses and wires. The best choice he saw was 90 degrees to one side, and he had to make a fast decision to take that. He was making a flat climb and had a lot of airspeed but, unfortunately, not a lot of altitude. The airspeed decayed rapidly after the loss of power, but he made the peanut butter and the bread come out even: He got the airplane into a clear area. There wasn't enough left of altitude or airspeed to flare and land, and the airplane was damaged. Nothing else was, so the outcome was a happy one.

Another pilot I know didn't do so well. Early one morning he hopped into a Cessna 140 to make a quick run to another airport. There was fuel in one tank and not in the other; unfortunately the fuel selector was on the dry tank. The engine quit at about 300 feet, and he turned back to the airport. As he hastened the turn and tried to slow the rate of descent, the logical thing happened. The airplane stalled. Fortunately, it was close to the ground and the impact wasn't catastrophic. The big scar on his head was always a reminder of the hazards of (1) taking off with the fuel selector on a dry tank and (2) trying to turn, power-off and close to the ground, and return to the airport.

Stalls and Spins

The episode of a return to the airport after an engine failure is a perfect lead-in to a discussion of stall/spin theory and recovery and the factors that influence a pilot to use technique poor enough to accidentally spin an airplane.

To begin with, an airplane in initial climb is operating at a speed relatively close to the stall and at a nose-high attitude. Using the Cardinal RG as an example, the best angle-of-climb speed of 67 knots is only 10 knots above an average power-off stalling speed with wings level. With 45 degrees of bank, which would probably be a minimum for someone hastening to return to the airport, the stalling speed in takeoff configuration (10 degrees of flaps) increases to about the 67-knot speed at the time of a power failure. So in case of engine failure it is absolutely necessary to lower the nose of the airplane and increase speed before doing anything else. Racking it into a turn first would only court disaster. Starting it into a turn while lowering the nose might be possible, but you'd be going into a critical maneuver.

Once the turn is entered, the forces at work are strongly against success. The airplane's power-off stalling speed is high, as is the rate of descent. Again using a Cardinal RG as an example, the rate of descent in a wings-level power-off glide is almost 800 feet per minute at 75 knots, the speed for maximum gliding range. This speed would not be safe in a 45-degree bank, as it wouldn't allow enough margin. Considering the airspeed and the increased rate of descent because of the bank (as well as that the airplane will be turning downwind, which could mean a rapidly increasing tailwind and resulting momentary decay in airspeed), you can see a few of the factors working against the pilot in this situation. At this low airspeed the turn might conceivably take as few as 20 or 30 seconds, but if the rate of descent is 1,500 feet per minute, it will take a lot of altitude.

Even more seriously, the visual sensation of a high rate of descent may cause the pilot to use too much back pressure on the elevator control, reducing the airspeed still further. The desire to get back to the airport may cause a similar inclination to steepen the turn, thus narrowing any margin between the actual and stalling airspeed and increasing the rate of descent even more.

Even though the nose of the airplane is down, the

wing of the airplane is flying at a high angle-of-attack (its angle in relation to the relative wind) and angle-of-attack, not attitude, is what counts. If the wing reaches the stalling angle, the result will be an even higher rate of descent and a lot of airplanes will enter a spin. If the airplane isn't stalled, there's still a very high rate of sink to deal with after rolling out of the turn and, power-off, it takes considerable airspeed to arrest that rate of sink. Ground contact probably starts becoming harmful to health at rates of descent in the 500 to 700 foot-per-minute range, and gets positively destructive at higher rates. If the airplane is stalled in the turn and a spin begins, then school is out. In the low altitudes involved in a turnback after an engine failure, there's just not room to recover from a spin entry.

It's clear that a turn after engine failure in initial climb should be explored at altitude, with a flight instructor. Just remember that if a 180-degree turn can be completed at altitude in, say, 500 feet, that doesn't mean it's safe to attempt a turnback when 500 feet above the ground. Turbulence, the effect of wind, and the pilot's tendency to steepen the bank and pull on the wheel at a high rate of descent can increase the altitude required to turn around safely. A pilot who has accomplished this in 500 feet at altitude might stall and spin while trying to do it from 750 feet after an actual power failure.

It's quite unnecessary to stall and spin a light airplane because of power failure. Consider two cases in which jet airliners lost all power and were landed more or less successfully, even though their crews had no training in power-off landings. The crew of a Southern Airways DC-9 lost both engines in a severe thunderstorm, made a power-off approach to a road, and maintained control of the aircraft through touchdown and until one wing hit an obstruction. There were survivors; if they had stalled the aircraft, it is not likely that there would have been any. The crew of a United DC-8 also had a complete power failure, yet they "landed" the big machine some miles from the airport and most of the people on board survived. Had they stalled the airplane, it would have been a different story. And if such highly wing-loaded airplanes can make a power-off glide and successful touchdown, it's certainly possible in a light airplane.

Crosswind

Let's move on to some variations on the takeoff.

A crosswind is the most commonly encountered situation that requires special technique on takeoff, and there are a couple of ways to deal with it.

If the runway is narrow, there's no choice but to track the centerline, making an absolutely straight roll. Aileron into the wind is used to keep the wings level in the takeoff, and directional control is maintained with the rudder. Some prefer to start the roll with full aileron and then reduce it as speed is gained and the ailerons become more effective; others just use aileron as necessary.

Any strong crosswind is going to be gusty, and the chances of making a bobble-free run are fairly small. As the effect of the wind changes, the heading of the airplane will vary slightly, and footwork will be required to keep it tracking the stripe. Both wind velocity and direction changes will affect the airplane. The key to success is in making prompt but smooth corrections to keep the airplane tracking as close to the center of the runway as possible.

Remember to let the airplane run in the attitude it assumes when properly trimmed for takeoff—no forward or back pressure on the elevator control during the roll. Add a few knots to the normal takeoff speed for crosswind operations, and when at this speed, rotate the airplane to a flying attitude without hesitation.

It's common for pilots not accustomed to a crosswind to have problems with directional control on takeoff and to try to solve this by forcing the airplane into the air. The result is often a trip through the rough, off to one side of the runway. Whatever the directional problem, it is easier to handle on the runway until the airplane is at a speed some knots above the normal liftoff speed. If liftoff speed is too slow, the aircraft might start drifting downwind and might touch back down after liftoff.

If the runway is wide, there's a variation on the crosswind takeoff that can be quite helpful. For example, consider a crosswind from the left. Start at the right side of the runway aiming at a point about 500 to 1,000 feet down the left side of the runway. A good number is about half of what you think the takeoff roll will be. This angle minimizes the crosswind in the first part of the run. As the airplane gains speed and nears the left side of the runway, the takeoff roll can be gently curved back toward runway heading. The centrifugal force of the turn will help keep the wing level. As in any crosswind takeoff, liftoff should be at a speed slightly above normal. I've made takeoffs in very strong crosswinds using this method, and it is effective. It should be practiced with an instructor familiar with the procedure before trying it on your own. Also, if the fuel tanks are not full, the procedure could cause an interruption of fuel to the engine if the turn forced fuel in the tank away from the line to the engine.

How much extra speed should be attained before liftoff in a crosswind? If there's information on wind velocity, and there are no gusts, five knots might be enough. If there are gusts, the spread between the steady and the peak wind might be a good value to add.

Soft-Field Takeoff

The procedures for soft-field takeoffs are well covered in training. Flaps as specified, full power, wheel back, get the nose up, and let the airplane run nose-high until it flies off. Then nose down gradually and accelerate to normal climb speed. The reason we do it this way is to get the nosewheel, usually the smallest of the three, out of the soft stuff and to progressively lighten the weight on the other wheels as the airplane accelerates. This minimizes the drag of the soft surface as much as possible and as soon as possible. Trouble is, in training we usually practice this on hard surfaces. Most pilots make their first real soft-field takeoff without the benefit of an instructor. When doing so, go back to the procedure of determining the halfway point on the runway and using that as a decision point. For best results, if the airplane won't fly at the halfway point, abort the takeoff. If the field is really soft, there should be no trouble stopping from this point.

A pitfall in the short-field takeoff is that under some

conditions an airplane might leave the ground but not be able to fly out of the ground cushion. The ground cushion is generally considered to extend upward about half the span of the airplane and in this area an interaction between the wing, the air, and the ground allows flight that wouldn't otherwise be possible. So if the airplane lifts off but then seems not to be accelerating, it might be telling you that it's not really going to fly.

Avoid the Hairy Way

If careful calculations are made for every takeoff, and conservative guidelines used, every takeoff should be without sweat. Do it without proper thought and preparation, though, and the situation can become sticky.

I recall a departure on which a combination of ingredients caused me to fall back on an anecdote related by a fellow who had flown heavily loaded B-29s to Japan in World War II. He explained that they would run almost to the end of the runway, where they knew it was time to lift off. Then going out over the island, they would aim the airplane to just barely clear the highest obstacle ahead, a water tank with a red and white checkerboard pattern. They reasoned that if the airplane would make it at all, it would make it when flown in that manner. From the way he said it, I felt there must have been some very close examinations of the tank.

I was with some folks one day who had one of the first of a brand-new twin on hand. They asked if I would like to fly it. Sure. The purpose of the trip was to take the airplane and a lot of materials to another airport.

The day was hot, and the runway only about 3,000 feet long. I asked a basic question about weight and runway length while boarding the airplane with their pilot, whom I considered in charge. He said that it should be okay.

We got the machine cranked up, and the air conditioning going, and I taxied the airplane into position for takeoff. The acceleration seemed okay at the start of the takeoff run, but the needle on the airspeed indicator still had a long way to go for comfort. We rolled and thundered past the halfway point of the runway in what seemed a lackadaisical manner, and a few moments later I asked the pilot if the engines were really developing the proper amount of power.

"I don't know, I'm not really that familiar with this airplane."

It had clearly become a case of being in the wrong place at the wrong time. The way things were going, it appeared that the airplane would indeed accelerate to a speed well above the stall before reaching the end of the runway. So having done all the unwise things to that point, I hoped that my judgment was now correct. I let it keep running on the ground until we were a few hundred feet from the end. Then I eased the airplane into the air and pointed it at the top of the tallest tree ahead. It roared out and over in good shape. My friend in the right seat never said a word, and if either engine had so much as belched, we'd have gone in. Years later, as I write this, I have just checked the pilot's operating handbook for that airplane and calculated the required runway length for that takeoff. It came to *precisely* the length of the runway. It was a lesson I'll not forget.

While the takeoff might not appear to be as challenging as the landing, it is a very important part of flying. Think about it this way: on a landing, the airplane is going progressively slower and the consequences of fouling up usually become less serious as speed decreases; on a takeoff, the speed is increasing and the reverse is true. The time to determine that a takeoff is going to be successful is in advance, not as the airplane fans the treetops in a hair-raising departure.

3

En Route Climb and Cruise

Once past the initial climb stage, going on up to cruise altitude becomes a matter of operating the airplane in the safest, most comfortable, and most efficient manner. The three don't always go together.

The most efficient way to climb to altitude is said to be at full power and the best rate-of-climb airspeed. Most engines are okay at continuous full-power operation, so there's no problem there. But the steepness of such a climb often compromises visibility in areas of possible conflict, and the noise and vibration levels at full power are often high enough to make such a climb uncomfortable.

Consider visibility first: The climb portion of the flight is conducted where there's a high likelihood of other traffic. Airplanes congregate around airports, and if you are flying from an airport in a metropolitan area of any size, the en route climb is in an area where airplanes are coming and going to a lot of different airports.

We often think of the steep climb as being bad for visibility because the nose of the airplane is so high. If the airplane is a powerful one, though, the airplane is not only pointed up, it is going up, so the blue sky in the windshield is reasonably representative of the airspace the airplane is about to use. To improve your tracking of airplanes that also are climbing, but in the opposite direction, some mild S-turns in the climb can enhance the view ahead.

More important to me is the potential of colliding with overtaking airplanes. When at best rate-of-climb speed in most light airplanes, we are flying in a manner that makes almost all traffic that is behind us and going in the same direction *overtaking* traffic. A Cessna Skyhawk flying level is probably overtaking a Cessna Centurion at its best rate-of-climb speed—barely, but still overtaking. Turn that around, and a Centurion at cruise will be overtaking a climbing Skyhawk at a rate of 70 or 80 knots. That's as much as 135 feet per second. If both airplanes are flying on the same heading, the pilot of the Centurion has to find the climbing Skyhawk before it becomes obscured by the nose of his airplane; the Skyhawk pilot has to look out the back window to find the Centurion, but in a climb attitude the view to the rear is limited. It is demanding. For the climbing pilot to do his bit on "see and be seen," S-turns would be in order, with a strong scan to the rear as well as in all other directions. The collision between the 727 and the Cessna 172 at San Diego was a classic case of a faster airplane overtaking a slower airplane that was climbing.

A cruise climb can give the advantage of better visibility and more comfort. In most airplanes, this means powering back to about 75 percent power and flying at a speed somewhat above that for the best rate of climb. It is less frenetic, too.

A Smooth, Hands-on Transition

On transitioning from the initial climb at full power and best angle-of-climb speed to a cruise climb, do things in a proper order to take maximum advantage of all your machine has to offer.

For example, the procedure in my Cardinal RG was to climb with 10 degrees of flaps, full power, and 67 knots indicated airspeed to 1,000 feet above the ground, then transition to a flaps-up 90-knot climb using 25 inches of manifold pressure (or full throttle above about 4,000 feet) and 2,500 rpm. The proper procedure for the transition was first to retract the flaps, to get rid of that drag. Next came acceleration to 90 knots. Finally the power was reduced. I watched a lot of people fly the airplane, and most would reduce power first, making the airplane struggle through acceleration and inducing a definite flat spot in the climb. I never quite understood why they did it that way. Surely it isn't logical.

On the Step

Another time to adjust power after doing everything else is when leveling off at cruise after climbing. There's always been a lot of talk about getting an airplane "on the step." What that means is accelerating from climb to cruise speed in a proper manner, one that accomplishes the acceleration as quickly as possible. I climb about 100 feet above the cruising level and then settle back to it, still at climb power, accelerating. When right on the altitude mark, and with the indicated airspeed up at the proper value, I then slowly reduce power to the cruise setting. Leveling right at the altitude and reducing power before the airplane accelerates can result in gradually reaching the airplane's true cruising potential.

The value of calculating and noting the cruise power setting during preflight is apparent at every level-off. Instead of looking through the handbook, or fooling with a computer, a glance at a note on the flight log sheet does it.

Navigation (Pilotage Still Works)

The most common modern-day method of navigation consists of going from VOR to VOR to VOR, letting the needle show the way. If DME is on board, it gives distance and groundspeed. This is a very simple form of navigation and there's nothing wrong with it, but some pilots invite trouble by not acquiring and maintaining proficiency at pilotage navigation—that done with a

map, a compass, and a clock. Even when navigating with
VOR and DME it's good to keep up with your precise
position in relation to airports and terrain. Flying mind-
lessly, staring at needles, isn't the best possible situation.
Navigation is like everything else in flying. Its finest form
is when everything is used, and in VFR operation, everything includes the old rudimentary method of flying with
finger on map, identifying towns by highway and railroad
patterns and drive-in picture shows.

The decline in ability at pilotage navigation is well
illustrated on hazy days in busy areas where pilots can
often be heard pleading for DF steers or for radar vectors to an airport. It's a sad state indeed when a pilot in
an airplane has to ask a person on the ground to assume
the navigating chore because he or she became mis-
placed in VFR conditions.

A few oft-neglected but basic principles form the
foundation of pilotage navigation. Time and heading are
equal partners in determining success. When you start
from an established position, note the time and fly a
consistent and correct heading. Passing over an identifiable position at a later time gives information that can be
used to calculate speed and track as accurately as with
VOR and DME.

If you should become slightly confused about position, using backstops can become a very important part
of pilotage. For example, you might decide to maintain
a heading of 270 degrees until reaching the Sus-
quehanna River. There you might decide to turn south
unless your precise position can be firmly established at
the river. There are numerous towns, bridges, and dams
along the river—all landmarks easily and positively identified.

An even better backstop is where a major highway or

railroad or river joins or crosses another easily identifiable artery. It's just like flying into a funnel; fly until one of these is intercepted, then follow it to the intersection of the two for a determination of your exact position. Combine VOR information for an extra backstop and it becomes even harder to get lost.

When in doubt it is important to avoid circling or aimlessly flying headings that are far from the computed heading. If you hold a heading and know how long you have flown that heading, it's a lot easier to find your way again.

Marginal VFR

Navigation becomes critical in marginal VFR conditions; that is, when the ceiling is at or below 3,000 feet and/or the visibility is less than five miles. That's in the daytime. (Everything changes at night, which will be covered in Chapter 10.) In day marginal conditions, the best VFR navigation often becomes "IFR," that is, I Follow Roads (or railroads or rivers). It's a lot easier to keep up with position when following something, and if the weather is the least bit grungy, salvation is often in knowing your precise position.

While position is one key to marginal VFR flying, respect for weather is another. In preflight planning, a minimum altitude for terrain and obstruction clearance was calculated. If VFR can't be maintained at that altitude, no question: Retreat, go elsewhere, go land, do whatever is necessary to remain in VFR conditions at the predetermined safe altitude.

One Step at a Time

In planning marginal VFR navigation, it's best to base the total plan on pilotage navigation. The VOR might be lost because of its line-of-sight limitations, and while an ADF can be very useful in certain situations, and as a backstop, it's better not to think of it as a precise navigational tool in conditions of limited visibility. Don't depend on the electronics, and take the flight one little bit at a time. Fly to a point, and if it looks good to go on to the next point, do so. If it appears the least bit doubtful, change the plan. It is true that VFR flying is legal down to one mile visibility when outside controlled airspace, but any pilot fond of his skin makes three miles the absolute minimum visibility.

Is there a way to judge visibility when flying? There's an old joke about the tower asking a pilot if flight visibility was three miles; the pilot answered "affirmative" and then turned to his copilot and said, "Yeah, a mile out front and a mile to each side, that's three miles." If a little high school geometry is used, flight visibility can be estimated more accurately and honestly than that. Simple triangulation is the way to do it.

In a 30-60-90 right triangle, the longest side is twice the length of the short side. When flying level, one short leg of the triangle is from your fanny to the ground. If you are 8,000 feet high, about a mile and a half, and can see a point on the ground that is an estimated 30 degrees below the horizon, or 60 degrees above a vertical line from the airplane to the ground, then you are seeing about three miles, or twice your altitude. That method isn't worth a lot at lower altitudes, because it's not possible to estimate angles much more finely than 30 degrees.

A good low-altitude method also involves triangulation. Sit in your airplane on the ground with the airplane at a level attitude and measure both your eye level from the ground and the distance to the closest point you see when looking directly over the nose. If, for example, your eye level is five feet off the ground and the point you see over the nose is 40 feet away, the ratio is eight to one. When flying level at 2,500 feet, about half a mile high, the visibility will be about four miles when you can see the ground clearly just above the nose of your airplane. It's far from an exact method, but it's better than a wild guess. And when flying along at 2,500 feet, any fuzziness of the view over the nose can be taken as a definite indication of troublesome weather.

When things start getting fuzzy, your preflight review helps keep trouble at arm's length. Trying to make detailed interpretations of terrain and obstruction heights while bumping along in marginal VFR weather is at best a tedious thing to do. When you are studying the map instead of looking outside, who knows what is going on? There might be another airplane or a mountain or a TV tower closing fast while you pore over the chart.

The ADF: Faithful Old Friend

It is always a good idea to avoid airways, and thus traffic, when flying VFR. The faithful old ADF can be a good aid when off airways. Often a much more direct route than the airways provide can be determined using the ADF and a combination of standard broadcast stations and nondirectional radiobeacons. (A list of standard broadcast stations is available with Jeppesen's J-Aid service.) If you combine pilotage with the ADF, more-or-less direct navigation in good weather becomes very easy. As in any form of navigation, keeping precise records of the time crossing over carefully and positively identified points is important.

Learning to use ADF is worthwhile whether you fly VFR or IFR. If ADF has any drawback, it is its affinity for alternator noise, or static. The alternator systems on light airplanes are, for the most part, adapted from automobile hardware, and unless properly filtered, they emit a lot of interference. The ADF is most susceptible to interference in the low band, where the NDBs broadcast, and if there's any audio hash at all with the ADF in the receiver mode, be wary of the relative bearing displayed in the ADF mode until you've had a chance to check it for accuracy.

Relax and Regret It

Most VFR cross-country is done in good weather, not marginal VFR. Here we have the choice of flying idly along or doing simple chores that will make life easier in case something goes awry. If there's an autopilot, most of us run it. If it will couple to the VOR, a lot of us do that, too. We hardly bother with flying at all. We may become pretty complacent.

I found out why it's worth keeping up with what's going on when flying southwest from Zanesville, Ohio, in a well-equipped Skyhawk I used to own. The airplane had area navigation equipment, I had set up a waypoint somewhere along the way, and I was flying along more or less mesmerized by the HSI. The distance was unrolling at an uncomfortably slow pace because of a headwind. If someone had asked my position, I would have said "Ohio." While I did know how far I was from the waypoint selected on the area navigation set, I didn't really know where that waypoint was in relation to anyplace other than a VOR station.

Then the four-cylinder engine commanded my immediate attention by running on three cylinders. I had never had one do precisely that before, and at first I thought the engine might be coming unglued internally. I had been having some trouble with lead fouling of plugs from using 100-octane fuel in an 80-octane engine, and started convincing myself that this, rather than a swallowed value or some equally unpleasant event, was the problem. Whatever the cause, it was clear that I would be landing very soon. I finally located the nearest airport, starting from scratch. I had first to determine my exact position on the chart, and then decide which airport was closest to that spot. I didn't have the visual chart opened, and had to find it. The airplane might not have held altitude indefinitely and I was suddenly very interested in the proximity of obstructions.

As it turned out, the closest airport was comfortably close, and I flew there and landed. But it was a lesson learned, and I now occupy my leisure time by keeping close tabs on the nearest airport.

Groundspeed: Staying One-up En Route

Keeping up with groundspeed has its rewards, too. My Cardinal RG was bereft of distance-measuring equipment, so I got in the practice of calculating groundspeeds. I missed having the precise distance and speed displayed at all times but got used to doing it on paper.

The toughest times to keep up with speed without DME are on the first leg of a trip, after a substantial change in heading when the wind is strong, and after flying through a front. In these three situations it's actually guesswork. I have found that the time en route on the first leg can be estimated with reasonable accuracy by calculating the time required at the estimated groundspeed and then adding a half a minute for each 1,000 feet of climb to cruising level. Conversely, when you pass the first point, subtract from the first-leg time a half-minute for each 1,000 feet of climb. Then calculate speed to have some idea of what it will be like on the next leg.

Another good en route occupation is checking weather every hour, or more often if some problem seems to exist. You can keep a running tab on weather by listening to the Flight Watch frequency or to a transcribed weather broadcast as you fly along.

Fuel Management

Some airplanes are blessed with a simple fuel system that is just "on." In others, we switch tanks at intervals and must keep up with how much fuel is left in each tank.

This isn't critical en route, when flying at a respectable altitude, but one item is worthy of consideration, as illustrated by an accident a number of years ago.

The airplane was droning along toward the destination and fuel in one tank was about to run low. When the pilot tried to switch, the valve handle came off in his hand. It had broken at the shaft. There was plenty of fuel in the airplane, but it was unavailable. The weather was at its worst with conditions below IFR minimums in widespread fog. The pilot hastened to try to make an instrument approach before all the fuel in the selected tank was exhausted. Unfortunately he wasn't successful.

The moral is to switch tanks often and not run any one tank down very low unless it's absolutely necessary.

There have been cases of fuel starvation when the pilot thought there was adequate fuel in the tanks and the clock and/or fuel gauges agreed. My father was involved in an incident that clearly illustrates this. His flight in a Cessna 180 was from Wichita to Columbus, Ohio, in the wintertime. The climb out of Wichita was in icing conditions up to 11,000 feet, where he was between layers. About an inch of ice accumulated on the airplane in the climb. No more ice accumulated, but that on the airplane remained; the temperature was below freezing at 11,000 feet. The fuel gauges stayed on full for over three hours, which seemed odd to him. The engine ran normally. When in the vicinity of Dayton, Ohio, the airplane's engine fell silent. What to do? Simple, tell the traffic controller what happened, extract the Dayton approach plate from the Jeppesen book and shoot a power-off instrument approach to that airport.

It all worked out well, and once on the ground he quickly found the cause of the power loss. The fuel vent,

which was on top of the wing of the 180 at that time
(1955), had iced over. This precluded the entry of air
into the tank as fuel was used. The airplane had rubber
bladder tanks, and these slowly collapsed as fuel was
drawn out. As they collapsed, they fed the engine normally but also squirted six gallons per hour overboard
through a small hole at the rear of the fuel vent. The
moral is to investigate any abnormal behavior of fuel
gauges. The fact that they remain on full is an indication
of a malfunction.

A pilot flying a twin had a fuel problem similar to but
not exactly like my father's.

He topped off his airplane and took off cross-country, only to notice the fuel gauges dropping at a precipitous rate. Thinking it was some malfunction of the gaug-
ing system, he continued. The gauges quickly reached
empty, and surprise, the engines ceased operating
shortly after the fuel gauges predicted they would. Why?
All the fuel had siphoned out because a fueling cap had
been left off. There was no big airport for this pilot to
use; he had to settle for a nice broad stretch of highway.

Leaning for Economy and Performance

The amount of fuel that goes through the pipe is
determined by the power setting and the mixture setting.
If operated full rich, most engines will go through a lot
of fuel in a short time; the endurance figures in the pilot's
operating handbook are usually based on the mixture
leaned when operating at or below 75 percent power.

There are different schools of thought on leaning.

Some pilots like to run their engines a little rich, to get maximum cruise performance and take advantage of the cooling properties of extra fuel. "Gas is cheaper than an overhaul," they say. Other pilots run with the mixture as lean as permissible. On unturbocharged Lycoming engines operated at 75 percent power or less, this is usually at a peak exhaust gas temperature setting, with an EGT gauge required to tell when it is at peak. On Continentals, maximum lean is usually from 25 to 75 degrees on the rich side of peak EGT, except at lower power settings —55 percent power or less—where peak EGT operation is permitted on some Continental engines. The recommendation in the pilot's operating handbook should always be followed.

The exhaust gas temperature gauge is the best indicator of mixture and the most precise tool to use in leaning properly. If you don't have one, it is still possible to lean with some degree of precision. There's a slight loss in power at a peak EGT setting. When flying an airplane with a fixed-pitch prop, leaning until there is a very slight decrease in rpm would indicate that the engine is operating somewhere near peak EGT. Lycoming approves of this procedure as long as the engine is running smoothly. With a constant-speed propeller, lean until you think you've heard a perceptible power drop, check the fuel flow or the position of the mixture control, and then abruptly enrich the mixture. If this results in a slight surge of the propeller rpm, then you know that the mixture selected resulted in a power loss. If it runs smoothly there, it might have been close to peak EGT and you might return to that setting.

For leaning to "best power" (the leanest possible mixture with no power loss) without EGT, the key point would be maximum rpm on a fixed-pitch prop. With a constant-speed, the mixture should be at the leanest point where an abrupt enriching doesn't result in a slight surge of the propeller.

Tales abound about the ways you can damage an engine by the misuse of mixture control, and it is true that an excessively lean mixture can hurt an engine, especially at high power settings. In an extreme case, engine failure can result. But leaning as per the manufacturer's recommendations shouldn't hurt an engine.

Stretching Gas

Pilots have been known to talk of "stretching" gas, as if it is made of rubber. This is really a reduction in power setting to achieve greater range. It works, but it is far from a new form of magic.

For example, imagine that you are three hours into a five-hour flight in a Cessna Centurion. If the airplane was fully fueled to begin with and was flown at high cruise per the book, 52.3 gallons would have been used at that point, leaving 36.7 gallons in the tanks. The true airspeed is 169 knots at a fuel flow of 15.3 gallons per hour, and the groundspeed 139 knots, reflecting a 30-knot headwind, instead of 15, which was forecast and used in planning. The distance remaining is 280 nautical miles. This would take just over two hours, and 31 gallons of fuel. The reserve on landing would be 6.7 gallons, which is far from enough. How about slowing down to a more efficient speed? Pull it back to 44 percent power for a true airspeed of 132 knots on 9.66 gallons per hour. The 280 nautical miles will take 2.75 hours at the 102-knot groundspeed, and fuel required will be 26.6 gallons. That would leave 10.1 gallons on landing, and indeed the fuel would have been "stretched" to allow landing with one hour's reserve. The question is whether the best option would be the 45 minutes additional flying time at the lower speed, or an extra stop. To some pilots, stopping is a good and restful break in a flight. To others, a stop is more fatiguing than continuing. In retrospect, as soon as the greater-than-anticipated headwind was experienced, adjusting the power and speed to afford steady progress to the destination would probably seem more sensible than going as fast as possible for three hours and then slowing to a relative crawl for the remainder of the flight.

Wind forecasts are often inaccurate, so recomputing fuel reserves, best altitudes, and power settings is an essential en route chore.

Choosing the "Optimum" Altitude

Sometimes the altitude selected before takeoff is not the best altitude en route, simply because there's no way to know what it's going to be like until you get up there. For example, I reviewed the weather forecasts and reports for a flight from Columbus, Georgia, to Trenton, New Jersey, and decided that 9,000 feet would be the best altitude. The only cloud cover along the way was given as 12,000 overcast. I found my 9,000-foot level to be in cloud, with some light icing. The overcast was lower than reported or forecast, and I changed to 11,-

000, which was on top of all clouds. With 12,000 overcast reported and forecast you wouldn't expect to be on top at eleven, but that's the way it worked out.

As much as for any other reason, we change altitudes en route to minimize headwinds. With a headwind, the old bromide is to fly lower because wind velocities are usually forecast to be lower at lower altitudes. This works to some extent but it is no cure-all.

On a westbound winter trip, the groundspeed was at 106 knots with the true airspeed at 144—painful at best. The cruising level was 6,000, IFR of necessity, and as soon as the area of clouds was left behind for the clear blue of a cold high-pressure area, the IFR was canceled. Go down low for less wind. It was too turbulent for comfort at a very low altitude, but 2,500 above the ground proved a nice blend of comfort and lighter winds. At least I thought the winds were lighter until I calculated the speed—100 knots, or six less than it had been at 6,000 feet. The true airspeed was five knots less at the lower altitude, and the wind was about the same. Earlier that day, when IFR, I had flown at 6,000, 8,000, and 10,000 to stay out of ice and found little difference in wind at those three levels, despite the forecast for increasing velocities with altitude.

Smooth Air and the Right Altitude

Which brings me to an observation about changing altitude to minimize the effect of en route winds: first, if you like comfort, flying low in turbulence is seldom worth any extra knots of groundspeed that might accrue.

So the usable altitudes are those where the air is smooth.
And in years of changing altitude seeking a better break
from the wind, I have found it to be a more or less futile
exercise. It seems that from the lowest level where the air
is smooth up to 12,000 to 15,000 feet, where there can
start to be some jetstream effect, the wind velocity does-
n't vary as much as the forecasts often suggest. The best
altitude is often the one at which the airplane operates
most efficiently. Certainly if the air is dead smooth in a
climb from 6,000 to 12,000, and no gradient wind effect
momentarily boosts airspeed and rate of climb, or makes
them sag, any velocity change is very gradual. And your
airplane will cruise a knot or a knot-and-a-half better for
each thousand feet you climb. Finally, the rocking and
rolling of turbulence does slow an airplane down a little,
which is another argument for staying in smooth air.

Traffic—See and Be Seen

En route is probably the time we become most lax in
looking for other traffic. Fly and fly and fly, look and look
and look—there's usually nothing there. When anything
is done for a long period of time without reward, the
tendency is to lose interest. And we soon lose interest in
looking for other traffic when none is spotted for a couple of hours. Then something will whiz by real close and
rekindle our vigilance.

I've seen some of the staunchest advocates of the
see-and-be-seen concept of traffic separation fly for long
periods while staring straight ahead. I've also seen them

spend long periods of time looking at charts or recording data with nary a glance outside. The simple fact is that the great expanse of airspace over this country and the relatively light traffic provide a better en route collision avoidance system than the see-and-be-seen system. This is proved by the relatively small number of en route collisions and the current lack of en route vigilance. It's safest to consider that this era is now ending and resolve to search aggressively for other traffic at all times when in visual meteorological conditions.

Don't use radar traffic advisories or controlled VFR in a terminal area as an excuse to quit looking for other airplanes. Traffic is very frequently not called by controllers, and in case of a system error only a pilot seeing another airplane can save the day. Any time the airplane is moving and in visual conditions, it's a good idea to constantly cover all the areas and work to minimize the blind spots of the airplane. If your airplane has a transponder, it should be on, and an altitude encoder makes things even better.

The value of vigilance, and of the transponder, was clearly illustrated by an en route collision between two single-engine airplanes, one IFR and one VFR. The IFR airplane was cruising along, level at 6,000 feet. The controller had called traffic a couple of times earlier in the flight, and the pilot had acknowledged but indicated that he didn't see the traffic. As the flight progressed, it passed over a VOR station and continued southwest. Though instrument rated, the pilot was relatively inexperienced in terms of total and recent flying, and the National Transportation Safety Board speculated: "His inexperience, the IFR operation, and the two previous advisories could combine to cause this pilot to believe that he would be provided further advisories before the other aircraft might be expected to become visible."

The other aircraft in this collision departed from a small airport southwest of the VOR and took up a heading toward the VOR, climbing. The flight was VFR and there was no evidence that the airplane's transponder was operating on the VFR code.

The weather was quite good as the IFR airplane droned southwestward, tracking away from the VOR. Meanwhile the VFR airplane climbed toward the northeast, probably tracking toward the VOR.

The primary radar capability (that which shows transponderless airplanes) in this area was virtually nonexistent. The commissioning flight check report on the radar site indicated that the only usable video from the site consisted of secondary (transponder-equipped) radar returns.

About one minute before the collision, both airplanes were in positions where each was visible to the other. Yet they continued. A ground witness saw them moving toward one another and did not see either take any evasive action. A primary target showed briefly on the controller's radar scope and appeared again 12 seconds later. The second appearance was assumed to be falling debris—wreckage of the two airplanes. The IFR airplane's transponder return had disappeared from the scope, and the pilot of another general aviation airplane in the area called the center and said: "I was glancing that way, the sky was clear, and all of a sudden there was just a black puff. It's right now at just about my ten, ten thirty position, just like a flak explosion from World War II, identical to it. . . ."

The airplanes had collided almost head-on. Lots of

"ifs" here. If either pilot had looked, he would have seen the other airplane. If the climbing VFR airplane had had an operating transponder, perhaps the controller would have seen it and given the aircraft as traffic, prompting the IFR pilot to see and avoid the airplane. If the VFR pilot had been using pilotage navigation and had not been climbing toward the VOR, the conflict with IFR traffic would not have occurred. Had the VFR pilot, lacking a transponder, called the controller and reported position, the two pilots would likely have been made aware of each other.

It is true that en route collisions are rare, but they do occur. And in virtually every case, there was ample opportunity for prevention.

A final en route item is preparing for the arrival. All available information about the weather and runway conditions at the destination should be gathered well in advance. While droning along level, charts can be consulted for frequencies. Just don't fixate on charts and neglect to scan for other traffic.

The predescent work should be taken care of in the same way as preflight activities. The objective is to leave the cruising level with all the necessary information and with as many things as possible set for landing. Avoiding surprises during the descent, approach, and landing is as important as avoiding surprises on takeoff and initial climb.

4

Descent and Landing

There are rules of thumb and formulas to use in planning a descent. The one I've found most useful calls for starting down five miles away for each 1,000 feet above airport elevation. If flying at 6,000 feet and landing at an airport with an elevation of 1,000 feet, start down 25 miles out, descend 1,000 feet per each five miles traveled, and reach the traffic pattern altitude (1,000 feet above the ground) five miles from the airport. At 150 knots, two-and-a-half nautical miles per minute, this calls for a descent rate of 500 feet per minute. Slower speeds result in lower descent rates—120 knots would be 400 feet per minute. Faster airplanes are often pressurized, so higher descent rates can be handled with comfort. It's easy to keep up with the progress of the descent. Any time you want to check progress, just make sure you still have five miles to go for each 1,000 feet above the airport. Of course, terrain might require a more rapid descent to some airports.

Pinpointing the spot at which the descent should begin is easy with DME, but it can be done quite accurately without it. A landmark or a VOR cross bearing is good enough. On the other hand, a guess almost never works.

A descent that starts late can foul up the arrival. When approaching an airport on the east coast one night I saw a late descent coming when the controller was unable to clear me to an altitude lower than 13,000 feet because of conflicting traffic. The 65-mile point, at which I would like to have started down, slid by; 60, 55, 50, 45. Finally at 30 miles the controller had it all worked out. I did my best on the descent, but it was obvious to pilot and controller alike that the airplane would never get down before the destination was reached. The result was some vectoring and a lot of badgering about increasing my rate of descent. The final approach course was intercepted a little high and fast, the balls were being juggled in haste, and I had to rely on the autopilot for more help than I like to accept in order to handle all the chores and keep the needles centered.

The descent objective should be to get the airplane down to an initial approach altitude, or traffic pattern altitude, a few miles in advance. Stabilize things there and run the prelanding check. When VFR it's much easier to find traffic in the pattern from pattern altitude than from a loftier perch, when looking down and having to sort airplanes out from houses, billboards, and shiny objects on the ground.

What about speed for a descent? Some like to maintain cruise power to get some extra speed, others like to reduce power and maintain the indicated cruise airspeed in the descent. The latter is more efficient, the former might save a minute or two. Either way, the mixture should be leaned in the descent much the same as it was for cruise—peak EGT if that is allowed, or a specified amount on the rich side of peak EGT. It will have to be adjusted as altitude is lost. Some pilots go to full rich for a descent, but this only serves to waste fuel and cool the engine more rapidly than necessary.

If the high-speed descent is chosen, it shouldn't be overdone to the point that the airspeed moves into the yellow arc. For my money, the airspeed should always be kept in the green. The yellow is a caution area, limited to smooth air only, and some turbulence is encountered on almost every descent. Who knows exactly where? As a matter of interest, calculations that determine the airspeed redline value on turboprop airplanes are similar to those used in determining the top of the green on light piston engine airplanes. That's one more reason always to fly in the green.

Engine-cooling Technique

Considerate care of an engine is in not cooling it too rapidly when descending. If the power setting is in the green for rpm and (if applicable) manifold pressure, this should be enough to keep the engine warm during descent except in extremely cold weather, when the power should be kept well into the green and the descent made very gradually. This is often difficult, as indicated airspeeds will be higher in very cold weather. Couple that with the need to keep the airspeed and all engine gauges in the green, and you may not descend rapidly enough. If the airplane is a retractable, extending the gear will help. Approach flaps might help, but use of approach flaps on a long descent should be limited to smooth air in airplanes where extension of flaps compromises the limit load factor. Gear or flaps, the limit speeds should be strictly observed.

Vigilance on Descents

Some of the more spectacular collisions have involved a descending aircraft. Where we can descend to traffic pattern altitude early to avoid looking for pattern traffic against a cluttered background, there's no way to avoid the cluttered background during en route descent. We have to revert to mild S-turns, which help make other traffic easier to see. When an airplane is on a collision course, it remains in a stationary position on your windshield or window. Turn a little and its position will change, adding some perspective, making the other aircraft easier to see. Keep looking out of the cockpit. Preparations for landing are best accomplished *before* leaving cruise and *after* leveling at pattern altitude.

Finding the Airport

Airports can hide. They can deceive, too. Some years ago there was a rash of air carrier landings at the wrong airport. The first I remember was a piston engine Convair that landed at Yazoo City instead of Jackson, Mississippi. The flight was inbound to Jackson from the north, at night, and the pilot mistook the well-lighted but poorly surfaced (muddy sod) airport at Yazoo for the one at Jackson. Someone said that the pilots realized at the last minute what was happening but opted to land rather than attempt a go-around.

Jets have landed at the incorrect airport, too. A DC-8 landed at Troutdale instead of Portland (Oregon) International, and a 707 once used the Ohio State University airport instead of Port Columbus. The airport restaurant at Ohio State was named the 707 Room in honor of this navigational error. None of these incidents was serious, but ones where pilots mistakenly used the wrong instrument approach plate—Springfield, Illinois, for Springfield, Missouri, for example—have had bad consequences.

Snow on the ground and haze make airports especially hard to find. One winter day when I was approaching Marion, Illinois, with another pilot, the visibility was perfect but deep snow cover hid everything. The other pilot, in the left seat, decided he had the airport in sight when we were about 20 miles out. He made a 30-degree turn to the right to head for it, but the airport should have been straight ahead if our navigation had been worth a candle. The map wasn't consulted, and the VOR on the field was ignored. No question, that is the airport.

The turn and the straying VOR needle raised doubts. There was a WAC chart handy, and it showed the airport northwest of town, with a big lake to the southwest. None of this corresponded with the alleged airport toward which we were heading. We went back to the VOR, which involved turning 30 degrees to the left, and found the town and lake—and airport.

If you expect that an airport is going to be obscured in haze or snow, an advance plan on how to find it should be included in the preflight study of the route. Random sightings don't work well. If the destination is changed en route, inflight work can establish the position of the airport in relation to other things and make it easier to see. Preflight planning can provide the means to find the airport using landmarks, roads, railroads, rivers, and navigational aids. The difficulties in finding an airport VFR are one reason why IFR flying is really easier. When IFR, there's a chart that gives precise directions to the airport.

Vigilance for obstructions can't be overemphasized when flying at low altitude, as when approaching an airport. They don't build TV towers in traffic patterns, but at some locations it seems there's a tall tower on downwind. Savannah, Georgia, and Hutchinson, Kansas, are two fields that come to mind.

Flying Hands-on in the Traffic Pattern

Have a mental picture of the airport and active runway before the airport is in sight. If it's an uncontrolled airport and there's no advance information on wind, or the wind is calm, circling will be necessary. Otherwise it will be a standard pattern entry at an uncontrolled field and as directed by air traffic control where there is a tower.

What is a standard traffic pattern entry? The rules specify only that in the pattern all turns are to be made to the left unless the runway is identified as one with a right-hand traffic pattern. Theoretically, straight-in approaches are legal, as are base, downwind, crosswind, and upwind leg pattern entries. Entering the downwind leg at a 45-degree angle is the most common method; this probably does maximize the view of other traffic, and it gives the advantage of fitting the airplane into the flow of traffic at a point where other aircraft should be flying level and at normal speed. Entering crosswind, you could fly up over an aircraft climbing after takeoff. Entering on base, you might encounter other traffic on a lower or higher base leg. A straight-in would not join the pattern until the airplane joins other aircraft turning final at different altitudes. There's no immediate conflict on an upwind leg entry, but there could be with climbing airplanes on the crosswind or downwind leg.

If the runway is crosswind, it will affect the traffic pattern. For example, landing to the south with a left-hand pattern and a southeast wind means the aircraft will be drifting toward the runway on downwind and will have a tailwind on base leg. If a proper correction for drift is made downwind, the turn to base and to final will be more than 180 degrees. This requires planning, and the downwind leg should be flown a bit farther away from the runway than usual. Otherwise there's a strong likelihood that the turn onto final approach will be overshot.

Overshooting the turn to final when there is a tail-wind on base leg is another classic setup for an accidental stall and low-level spin entry. Power is usually off or at a low setting. The tailwind on base gives a visual illusion of speed that is not borne out on the airspeed indicator. As the pilot sees he is overshooting the turn to final, the temptation is to tighten the turn, increasing the stalling speed of the aircraft. Some might even push a little rudder to hasten the turn and use a lot of aileron against the turn to combat an overbanking tendency. If an airplane stalls in this configuration, most will break out of the bottom of the turn and start to spin to the left. At these altitudes, there's never room for a spin recovery.

If you have to tighten the turn to the point that G forces are felt, or if it requires more than slight aileron against the turn to keep the bank from steepening, it could degenerate into a loss of control. Time to level up, go around, and come back to try again.

The Sore Spots

Back to traffic at uncontrolled airports for a moment: If all pilots monitor the appropriate Unicom frequency and make position reports in the pattern, it provides a picture of what's going on. But there's no guarantee that there won't be pilots out there who are not talking or listening, and vigilance for traffic is both primary and crucial. Most collisions occur on final, but there are plenty of other hairy places in the pattern and near the airport. All airplanes have blind spots, and as you might expect, there are numerous cases of low-wing airplanes settling down atop high-wing airplanes on final.

Be especially wary when you depart from normal pattern procedures. For example, when making a steep approach, scan the airspace below, where someone might be making a shallow approach. And when making a shallow approach, eyeball the airspace above. When flying close-in pattern, watch for someone on a long final. By doing it differently, you might well be flying where other pilots won't think to look. Thus the see-and-avoid burden falls more on you.

An instructor was shooting landings to the northwest with a student on a short runway. The airport also had a north/south runway. The two didn't cross; rather, the one to the northwest began a couple of hundred feet beyond the north/south runway. It was a quiet afternoon, with no other traffic—probably not another airplane within fifty miles, or so the instructor thought.

He was slightly in error; there was at least one other airplane within fifty miles. It was to the south, inbound, and its pilot had made the decision to land straight in, to the north. Neither airplane was radio-equipped (this was a long time ago).

The instructor and student kept droning around the pattern as the other aircraft approached the field. The student was doing well and would be ready for solo soon. The airplane was a Cub, and the instructor was leaning forward, talking to the student about the upcoming flare for landing when he saw an airplane to the left on a straight-in approach to the north runway. In a split second the airplanes merged and then separated. The student and instructor landed okay, and while the other airplane was wiped out, its pilot was only bruised. Three people learned a lesson they would never forget: When operating on an airport with multiple runways, look for traffic on all runways and never cross one without first looking.

This also applies to controlled airports. I was on final one day and noticed another aircraft making an approach to a crossing runway. It looked as if we'd be close at the intersection of the runways, so I checked the tower. They said there were no other aircraft in the area and that I was cleared to land. Tower was only partially correct. No other airplane was in contact with them, but a pilot whose airplane had had an electrical failure was in fact making an approach to the other runway. It pays to make sure for yourself, even if there is a control tower.

Concentration on Approach

Vigilance is needed most at the most challenging times in flying. A good landing follows a good approach, and this good approach must be planned and executed while paying proper attention to the surrounding airspace. You can help by making all approaches as much alike as possible. Consistency in approaches usually helps landings. There are times when we have to vary the approach—at a big and busy airport where you might be urged to keep the speed up on final, for example—but at some point every approach must reach the correct speed and configuration.

The Navy Way

The Navy makes interesting approaches in daytime visual conditions. The objective is to be consistent and to keep traffic close to the carrier. Final approaches are very short. In fact, Navy pilots probably average shorter finals in F-14s than most flight schools do in Cessna 152s or Piper Tomahawks.

How do they fly the downwind at a consistent distance when there are no ground reference points? Simple. It's always at the same altitude and they can use a reference point on the wing (with the wings level) in relation to the ship. (On a high-wing airplane, a position on the wing strut would do. No strut, do your best.)

The airplane is set up for landing when passing opposite the approach end of the runway on downwind. The gear and flaps have been extended at that point, and the descent to landing begins. So does the turn toward the ship. The carrier is moving, so starting the turn opposite the landing point still leaves some room for final approach.

The Navy has visual approach slope indicating equipment, but plain eyeballing still plays a role as the airplane progresses toward the carrier. Most pilots know that when you are flying down a slope, the point on the ground toward which the airplane is tracking remains in a stationary spot on the windshield. This remains true if you change speed and rate of descent; then a new point remains stationary. It's easy to see on final, but it can also be used on base and when turning final. It's not a stationary point in the windshield at this time; it is the angle down to the landing point. Only with practice and expe-

rience can we learn how the runway is supposed to look
on downwind, base, and final.

The Power/Pitch Controversy

On final, if the airplane is a bit high, what to do? The
Navy adjusts altitude with power and airspeed (or attitude) with the elevator control. Is this a logical way to do
it in light airplanes?

The debate about power versus elevator to control
airspeed or altitude has raged for years. It has provided
fodder for the typewriters of aviation journalists (this
one included), and it has totally confused the FAA, which
has a lot to say about flying technique. At one point the
FAA adamantly insisted that the elevator controls altitude and the power airspeed, and then it did an about-
face and insisted that the opposite is true. Then it hesi-
tated and all but admitted that there was a difference of
opinion. There should be difference of opinion, because
the subject is broad and not subject to ironclad rules.

In light airplanes it is important to remember that
when the chips are down and things are looking bad, the
elevator control should be considered the primary tool.
Airplanes stall because of an excessively high angle-of-
attack, and the elevator does the quickest work on that
number. If the airspeed appears low or is deteriorating
toward a low figure, or if control of the airplane feels in
doubt, no question, reduce the angle-of-attack with the
elevator. Put the nose down. There have probably been
more people killed in light airplanes because of failure
to manage angle-of-attack properly than for any other

reason; the most often misused control is the elevator.

What is the best way to manage the elevator and power control to get the airplane to the runway on a normal approach and to have it in the best possible configuration for landing once it gets there? First, recognize that the objective is to get the whole airplane, not just the altimeter and airspeed indicator, to the runway. To arrive gracefully, a pilot has to think in terms of the total picture, not just in terms of controlling instruments.

Let's say we are on a half-mile final, a rather short final by civil aviation standards. The proper airspeed for the approach is 75 knots, but the indicator is on 80. The touchdown zone of the runway is moving lower in the windshield, indicating that the airplane is headed for a point beyond. Two things need attention. I suppose that the proponents of "this controls that" would suggest that such a situation be taken apart and done in separate steps. A good pilot would reduce power and lower the nose simultaneously. Then he'd quickly check to see if the new attitude and power setting were achieving the dual purpose of steepening the glidepath and bringing the airspeed back to the desired value.

These relationships and precise values can be clearly seen in smooth air. But most approaches are made in turbulence. Here airspeed and altitude are affected by updrafts, downdrafts, thermals, and gradient wind effects (shear). Here the test tube approach crumbles even further. You had better keep your eyes open and the seat of your pants sensitive, because what you see and feel is very necessary to reaching the end of the runway.

A Piper Pacer I once owned was a real teacher of lessons on approaches. The one I remember best was at

a little field on Lake Winnipesaukee in New Hampshire.
As we circled, my wife said that it looked more like a track
field than an airport. Indeed, the strip was about 1,500
feet long.

I don't remember the speed used for an approach in
the Pacer, but it was probably about 55 knots or a bit less.
There were trees at the end of the runway (and all
around the airport) so I set up a rather steep approach
slope that would clear the trees and get the airplane to
the runway as soon as possible. For my money, a consistent steep approach all the way in works better than coming over the trees, reducing power to increase the descent rate to the runway, and then trying to make a
decent landing. Set it up at the proper speed and a good
rate of descent, fly it over the trees and right to the flare
height, then land.

This worked beautifully in the Pacer. The airplane
didn't have a lot of span, and with the flaps down and the
airspeed at a proper value, the rate of sink was good. The
end of the sod runway was moving steadily toward me,
in the correct spot on the windshield.

As the trees were cleared, my eyes and the seat of my
pants told me very quickly that things had changed. The
rate of sink increased dramatically. I did not even look
at the airspeed indicator, but it was surely dropping.
There was only one thing to do—try to control altitude
with power. To increase angle-of-attack by trying to control altitude with elevators would have only accelerated
the Pacer's sinking spell.

The power worked fine, though there sure wasn't an
overabundance of it. The throttle was all the way in when
the airplane touched the ground, tail low. It had taken
the full authority of a 125-horsepower engine to handle

the situation. The touchdown was reasonable, and when the airplane touched I pulled the throttle back and rolled to a rather short stop. It probably looked like a fine approach to a deaf person standing and watching, but anyone else would have known that a lot of last-minute effort was necessary to keep from spreading the gear at touchdown.

This was a classic case of wind shear on approach. There was a headwind of maybe 15 knots above the tree line and virtually none below it. When the 15-knot headwind was abruptly lost, the airspeed decayed. The airplane wanted to nose down to regain airspeed, and the sink rate increased. It happens every time with a rapidly decreasing headwind. (The opposite occurs with a rapidly decreasing tailwind.)

To deal with these things, the pilot has to respond logically—not mechanically—to the needs of the moment. On that approach, the only salvation was in a massive dose of power.

How Fast Is Fast Enough?

What is a good approach speed? That listed in the book is usually about 1.3 times the stalling speed for the particular amount of flaps selected. That should be adjusted upward a bit for turbulence or wind gusts—some pilots add the spread between the steady wind and the peak gust (15 gusting to 30 would add 15 knots to the approach speed), and others add half that amount. Most who purse their lips and mutter in parables about precise amounts to add for gusts have never really flown a light airplane in strong winds. The airspeed jumps wildly. Keeping a Cessna 152 nailed precisely to 68 knots indicated on a day with gusts to 30 just doesn't work. We might psych ourselves up to do this, note the fluctuations, and then add too much speed; the result is arrival over the runway with excess speed and a difficult landing.

One way I've found to compensate for gusts and turbulence is to select an attitude and power setting that takes the airplane toward touchdown while keeping the airspeed roughly equal to the bottom limit of the recommended approach speed. If, for example, the normal approach speed is 65, I'm happy if the airspeed is fluctuating in a range from 65 to 80, or whatever, as long as the 65 isn't violated. If it seems to be getting close, I run with the nose a bit lower and the power setting a bit higher. This way I'm not bothering with the total fluctuation, just with one end—the end that relates to a margin above stalling speed. In extreme turbulence I might raise this lower limit a little above the normal approach speed.

General aviation pilots don't usually make approaches as close as the Navy does, but there's always some discussion about where in the landing pattern the airplane should be in the landing configuration and at proper approach speed. Some would have it configured and the airspeed nailed miles from the airport; others reach the target speed just before crossing the fence.

I saw a beautiful example of doing it late when flying with a really talented pilot in a Cessna Citation. We were landing at the Cessna strip on the east side of Wichita, and because of traffic at adjacent McConnell Air Force Base we crossed the north/south strip headed east, about 4,000 feet above the ground. A mile to the east, the pilot started a gradual arc to the left. At the start of the turn the airspeed was on 250 knots. As the turn progressed, he gradually reduced speed. Flaps were extended, speed was further reduced, the gear was extended, and when he rolled out on final about 1,200 feet from the end of the runway the airspeed had just reached the target airspeed for final approach. It had been a beautifully executed approach, continuously descending, turning, and slowing right up to that perfectly stabilized short final.

Oh, but that we could all fly so well.

For most of us flying light airplanes it's best to routinely plan on one mile of flight before touchdown with the airplane in approach configuration and the airspeed stabilized at 1.3 times the stalling speed, or whatever is listed as a normal approach speed in the pilot's operating handbook. Speed might be allowed to decay to 1.2 times stalling speed when crossing the airport boundary at a proper altitude, followed by the flare for landing. If we do this every time, landings will improve, because each will start from the same point—crossing the end of the runway about 50 feet high with the airspeed at from 1.2 to 1.3 times stalling speed.

Expedited approaches—where the speed is kept as high as possible for as long as possible—take practice. Regardless, the object is always to cross the end of the runway at the proper speed and in the landing configuration.

The Flaps Flap

There's long been a flap about the use of flaps on approach and landing. Some say that full flaps should be used except in a crosswind. Others suggest that flaps be used as necessary.

First, consider that the function of wing flaps is to increase both lift and drag. But this is not accomplished evenly over the travel of the flaps. Increasing lift means lowering the stalling speed, and the pilot's operating handbook tells the tale on this. Look at a Skylane RG as an example: There's a three-knot decrease in stalling speed while moving from a no-flaps configuration to 20-degree flaps. There's virtually no further decrease in stalling speed from 20- to 40-degree flaps. There is a lot of drag in going from 20 to 40 degrees though, as any Skylane pilot knows. There the increase in lift is in the first part of the flaps travel; the rest is drag. A Centurion is somewhat the reverse, with but a one-knot drop in stalling speed in the first 10 degrees and an eight-knot drop in going from 10 to 30 degrees. In a Navajo Chieftain stall speed decrease is rather evenly divided over its 40-degree flaps travel, and a Mooney 201 gets almost all its decrease in the first 15 of its 33 degrees of travel. The way flaps generate lift depends on the particular design.

There is also a substantial difference in the way flaps generate drag. On a Tiger or Cheetah the flaps generate a mild amount of drag, on a Skyhawk they generate a lot.

The effect of flaps varies so much from airplane to airplane that it seems unwise to have a firm rule for their use. Besides the obvious stall speed and drag production, flaps affect the handling qualities on an airplane.

It can be argued that any change in handling qualities with flaps is indirect, but anything that requires a change in flying technique affects handling qualities, and changes in technique are required with various settings of the flaps.

If the extension of flaps causes a pitch change, this can be quickly handled with the trim. The increase in drag is used to adjust the glide path. The lower stalling speed is used to minimize touchdown speed. The flaps do things in return for these favors. They change the nose-up attitude at which the airplane stalls, because extending the flaps effectively increases the camber of a portion of the wing and reduces the stalling angle-of-attack. Although we can land with the nose quite high on a tricycle with less-than-full flaps, landing with them all the way down often results in a near-level touchdown even with the elevator control full aft. If the longitudinal control system of the airplane includes downsprings—which are used to enhance stability at aft cg loading—the pilot will have to pull harder for a tail-low touchdown than would be necessary with partial flaps.

The downwash of air behind the flaps can also reduce elevator effectiveness; this is more pronounced on some airplanes than others. T-tails seem to be exempt.

Finally, the lower stalling speed with full flaps means that the ailerons and rudder will be less effective at touchdown—simply because of the lower airspeed.

If a theoretical advantage can be found for using full flaps on every approach and landing, it is in consistency. Each landing could be the same. But this is true only if the center of gravity is in the same place for each landing, because, like flaps, cg affects the longitudinal handling qualities of the airplane. With the cg aft, the elevator

power is adequate to get the nose higher on landing, and
the touchdown speed will actually be a little lower. When
it is considered that on many airplanes the stall speed
decrease is concentrated in the first part of the flap
travel, and this is related to all the other things mentioned, it is clear that the advantages of full flaps are
often in areas other than low touchdown speeds.

Steep approaches are better than shallow approaches. The things that airplanes crash into (obstructions or the ground) stay farther away longer when the
airplane is on a steep approach than when it is on a
shallow, dragged-in approach. The options in case of
mechanical trouble are better on a steep approach. And
with maximum drag from full flaps, a steep approach can
be flown with enough power to keep the engine from
cooling rapidly. In light airplanes, the steep, maximum
drag approach also means less noise for the neighbors.
Flaps are a key to steep approaches.

Properly used and understood, flaps are valuable
tools; an exception is crosswind landings. Some light
airplanes are very awkward in a crosswind with full flaps,
and most are much easier to land with partial or no flaps.
A nose-high touchdown at a higher speed is possible,
and aileron and rudder control will be more effective for
dealing with the crosswind. But the approach will be
shallow—very shallow on some light airplanes—and the
landing will require a lot more runway than with full
flaps.

If you think there's no valid conclusion to this argument, you are quite correct. The subject of flaps is too
broad and should really be taken on a case-by-case basis.
When I had a Skylane, for example, I always used 30-
instead of 40-degree flaps for landing when there were
no rear passengers. The airplane was easier to land in

that configuration. But in my Cherokee Six, Skyhawk, and Cardinal RG it really didn't matter—they were easy to land tail-low, even with forward loading and full flaps. The Cardinal RG was pretty good in a crosswind with full flaps, but not the Cherokee and Skyhawk and Skylane.

The Skyhawk presented a special challenge because it was a good glider with no flaps and the approach was very flat. The slip became a very important part of the flaps-up approach. The main point on flaps is that what's good for one airplane might not be good for another, so beware all-encompassing pronouncements. Make your decision based on the airplane being flown.

Too High, Too Fast?

On any approach the pilot should ask himself "Is this approach going to lead to a normal landing on the first one-third of the runway?" Honest answers well in advance can avoid real problems later on.

I was on an airline flight, a Boeing 727, and was watching the gusty-day approach with interest. The pilot had his choice of landing to the north or landing to the west. It seemed that he started off with every intention of landing to the north and then switched to a rather close left base to runway 27. Final was turned at what seemed a slightly high altitude (even from row 21) and, shortly after turning final, the pilot instituted a go-around. When we were at the gate a few minutes later, I asked the captain why they had gone around. He smiled sheepishly and said, "The approach wasn't stabilized. We were just too high."

The same type airplane was involved in a serious

accident when the pilot flew the approach a little fast at a marginal-length runway and decided too late to go around. Then he changed his mind and tried to complete the landing. The airplane went off the end of the runway and was destroyed. Investigation revealed that the approach speed was high and the flaps were not fully extended. Excessive floating on the landing resulted from a combination of these things, plus shifting and gusty surface winds.

Go-arounds are good for both body and soul; they should be instituted the moment there is any doubt about an approach leading to a normal landing.

Technique: Safe Go-arounds

It is apparent that good go-around technique is early decision. The later the decision is made, the more the potential hazard.

Flying light airplanes we don't often overshoot large airports. It's the small strip that usually catches us, and a late go-around can be as touchy as a takeoff from a strip that is too short. True, the landing airplane starts a go-around with some speed, and perhaps even some altitude, but it is usually in a maximum-drag configuration, flaps fully extended, and there is a lot to do on any go-around. Taking off, everything is set at the start of the roll; that advantage is lost on landings.

Flaps are the most significant source of drag on most airplanes, and they are the first things to work with when a go-around starts. Retracting flaps does a number of things to change an airplane, and there's more to getting them up than merely flipping a switch.

A student pilot learned this the hard way while trying to land on a 1,900-foot strip. His approach was high and fast, and the airplane didn't get close to the surface until about half the runway was behind. The student flared, the airplane floated, and the trees off the end of the runway were looming. The airplane was over the last third of the runway, which had a downhill slope, and it still wasn't on the ground. There was a creek at the end of the runway, and the student made the decision to go around—too late.

Power was applied, and the airplane started struggling upward, flaps still fully extended. There's no way to know what the airspeed was, but it's bound to have been too slow. As the airplane approached the trees, the pilot retracted the flaps. The nose of the airplane was seen to pitch up a bit; then the airplane stalled and crashed into the trees.

Without a doubt, the decision to go around was made late, but a crash may have been avoided had the flaps been managed properly. Clearing the trees was his objective, so one might assume that the student thought getting rid of all the flaps would make things better. The reverse was true.

Retracting the flaps on this particular high-wing airplane resulted in the nose pitching up a bit. There would also be an increase of about six knots in the stalling speed. The combination of an out-of-trim airplane, a higher stalling speed, and a visual illusion (from the looming trees) that the airplane needed to go up, was too much. If the flaps had been partially instead of fully retracted, the airplane might have been capable of flying over the trees.

When flying an airplane with the majority of stalling speed decrease in the first of the flaps travel, it's obvious that the flaps should not be fully retracted in any critical situation. Instead, the position that results in the least drag and the most lift should be selected. The airplane will probably be out of trim, and it's up to the pilot to use whatever control force is necessary to maintain a safe pitch attitude. If it's a retractable, the gear should be retracted as soon as the flaps are adjusted unless the pilot's operating handbook suggests otherwise. (On some airplanes it's said that gear retraction should be delayed until all obstacles are cleared. This would be a result of there being less drag from an extended gear than from one in transit.)

We don't make many go-arounds and they should not be neglected when checking out in a new airplane. The maneuver need not be critical; just start early.

Technique: a "Normal" Landing

When we get an approach just right, the landing can become a beautiful maneuver. As the airplane wafts over the numbers, moving the elevator (or stabilator) control aft results in a smooth and continuous increase in nose-up attitude and an equally smooth decrease in the rate of descent as airspeed is sacrificed to reduce the sink rate of the airplane. The pilot has a good mental picture of the proper nose-up landing attitude, and the airplane reaches this attitude at the precise moment the last of the altitude is lost, resulting in a gentle kiss of the rear wheels on the runway. The airplane is allowed to roll along on the rear wheels for a bit; then the nosewheel is gently lowered. (The landing in a taildragger is a little different. *See* page 104.)

But all landings don't work that way, and if there is one most common reason for landing difficulties, it has to be excessive speed on final approach. The most difficult place to get rid of speed in an airplane is in the ground cushion, which is the airspace extending from the ground up to an altitude equal to half the wingspan of the airplane. The interaction between the wing and the air and the ground makes it want to keep flying, or said another way, to decelerate more slowly. In fact, an airplane decelerates about half as fast in the ground cushion as out of it. This is a prime cause of the long "float" that follows an approach at higher-than-recommended speed. Knowledge of this can be used to minimize the effects of a fast approach. All you need is to get rid of any extra speed before the airplane is in the ground cushion.

If the airplane is flown into the ground cushion fast and starts to float, there's a strong temptation to try to touch down before the airplane is in a proper landing attitude. This is unfortunate, because the airplane is not designed to be landed in a level pitch attitude, much less nosewheel first. A lot of nosewheels are broken off following a bout with an attempted level landing.

When an airplane is going faster than normal in the ground cushion and the elevator control is moved forward to hasten contact with the ground, the result is usually a bounce. If the pilot's motive was to put the airplane on the ground by moving the wheel forward, the instinctive response to the little bounce might be to try again to fly the airplane onto the ground by moving the elevator control forward. The bounces can increase in intensity, along with the pilot's responses to the bouncing, and you can see why the nosewheel might eventually fail.

When flying tricycle-gear light airplanes, bear in mind that moving the wheel forward has no place in a normal landing once the airplane is in the ground cushion and the landing has started. Moving the wheel forward is counter to what needs to be done. If the airplane balloons a little because of excessive speed or too rapid movement of the elevator control aft, don't do anything with that control until the airplane starts to settle again. Then resume trading airspeed for a reduction in the sink rate. If the sink rate seems excessive, a little power can help.

This theory of landing is sometimes disputed in turbulent and gusty conditions. For one thing, we may have added to the approach speed because of gusts; if it is really wild, it might not be advisable or possible to dissipate this excess airspeed while above the ground cushion. For another thing, the airspeed can fluctuate once the airplane is in the ground cushion, and ballooning can result. Anyone who has flown much has run into the situation where the airplane seems to develop a mind of its own over the runway—even after an approach at what was considered a proper speed.

Despite the vagaries of turbulence, the principles of landing remain the same. The objective is to have the airplane touch in a proper landing attitude. The dangers of forward elevator control to make the airplane land are even greater because turbulence could aggravate the hit or the overcontrolling that might follow a bounce. Think back to your last airline ride on a Boeing 727. The way most lads land those beasts is to get close to the ground, get in a landing attitude, and let the airplane land. You don't see (or feel) the gyrations that we sometimes go through landing on a turbulent day. They don't always make smooth landings, but they don't often bounce.

Landing attitude is a key, and flaps play a significant role here. With no flaps, a light airplane might well be near a landing attitude when crossing the fence at 1.3 times the stalling speed; with full flaps the airplane is quite far (in a nose-down sense) from the landing attitude. Consider how much easier it is when the airplane is near the landing attitude as it crosses the fence; all that's left to do is adjust attitude a little and lessen the rate of sink just before touchdown. That is a good argument for making flaps-up approaches when it's wild, windy, and turbulent. But the decision to do this has to be made with full consideration given to the extra runway length required because of higher approach airspeed and the shallow approach that will result in many airplanes.

A form of "cheating with flaps" is rather common. Use full flaps for a steep approach, regardless of wind and turbulence. When the airplane is in the ground cushion, and the landing has started, slowly retract the flaps and increase the rate of up-elevator application as necessary. Raising the flaps increases the stalling speed and greatly facilitates getting the airplane onto the ground. If done properly, the landing comes very soon after the flaps start retracting.

I once used a very small airport in a Twin Comanche, an airplane that is prone to float. I used the technique of retracting flaps at this field. Among other things, one has to get the flaps up on a Twin Comanche for the brakes to be effective, and brakes were important at this airport.

My drill was to come over the fence at 70 knots, start the flare, hit the flap switch, land tail-low, get on the brakes, and stop with some room to spare. In retrospect, I probably should have selected a larger airport. But it worked many times for me.

The drawbacks to retracting flaps in the flare are rather obvious. If it is done with much altitude, the result could be a dropped-in landing from high enough to bend something. If flying a retractable and by accident you hit the landing gear switch instead of the flap switch, the results are embarrassing. Too, the rate of flap retraction is slow on some airplanes and fast on others, so this technique has no across-the-board application.

Technique: Aerodynamic Braking

What should be done about the flaps after a normal landing? Some pilots retract them immediately, others wait until off the runway.

In favor of retracting flaps immediately after touchdown is better braking on some airplanes. The airflow over the flaps makes the rear wheels lighter on the ground while rolling at good speed, and the result can be ineffective braking, tire sliding, and at the extreme, a blowout caused by a locked wheel. Against retracting is the aforesaid trick of grabbing the wrong handle and pulling the landing gear from under a retractable. It is true that retractables have micro switches to preclude gear retraction on the ground, but these are often ineffective when there's not much weight on the wheels.

Also, there is some aerodynamic braking available from extended flaps.

If the flaps are left down after the landing and the elevator control is held full aft, the airplane will naturally slow down quicker than when rolling along level. There is good weight distribution on the wheels and the brakes should be reasonably effective on most airplanes. Some pilots hold the wheel full back to keep the nosewheel off for as long as possible and maximize aerodynamic braking. This is okay, but if the nose obstructs the view of the runway ahead on rollout, it could be counterproductive.

The procedure I prefer is to pick runways that are not of a critical length and leave everything alone on the landing rollout. Once off the runway I retract the flaps, open the cowl flaps, and do whatever else is necessary.

Technique: Crosswind Landings

The landing process becomes more interesting with a crosswind, but the condition shouldn't be intimidating. A crosswind should be handled as a matter of second nature. If we use the wing-down method in a crosswind, all we are doing is slipping into the wind at a rate equal to the velocity of the crosswind component. The wing is lowered as necessary to take out the drift, and the rudder is used to keep the nose of the airplane pointed straight down the runway. It's simple.

The other way of handling a crosswind landing is to crab as necessary to keep the airplane tracking down the runway and then remove the crab with rudder and opposite aileron just before the airplane touches. To me, this is the more difficult method, as the pilot must know precisely when the airplane is going to touch down. That's not always easy to judge in turbulent conditions.

Taildraggers

Everything said about approaches and going around applies equally to tricycle and tailwheel airplanes. It's only in the touchdown that different techniques are used.

There are two distinct and very different ways to land a taildragger. A three-point, full-stall landing is the most commonly used. The airplane is flared and held off the runway until the elevator control is full aft. The airplane should be in a three-point attitude, and if it all comes together when the airplane is but a few inches above the runway, the touchdown is smooth.

A wheel landing, on the main wheels only, defies the principles of proper landing technique in a tricycle, as well as those of a three-point in a taildragger. The airplane is flown onto the runway level (and very gently); at the moment of contact the elevator control is moved forward slightly to make sure the airplane stays on the ground. Usually the approach is at a slightly higher-than-normal speed to insure touchdown in a level attitude.

Practice Makes Perfect

During the time I instructed, most flying was done in taildraggers. Regardless of the landing gear configuration, I never felt that I "taught" anyone how to land an airplane. The instructor's role seemed to be one of a demonstrator, explainer, answerer of questions, and protector of the airframe while the student experimented and *taught himself* how to land the airplane. Landing defies description. If you tell someone how high to flare, there is really no way to know that the picture is forming in his mind as intended. The landing can be shown to a person, but seeing it without feeling the required pressures in the controls is only half a loaf. It certainly doesn't do a lot of good to "follow through" on the controls, because experiencing the motions without feeling the pressures is still inadequate. No, to learn to land an airplane, you have to practice landing the airplane. This is true whether you are learning to fly or checking out in a new airplane.

5
High-Performance Singles

The FAA defines a high-performance airplane as one with more than 200 horsepower or with a retractable landing gear, flaps, and a controllable propeller. It's best, however, to expand that definition and think of a high-performance airplane as one that exceeds the complexity or performance capability of the last airplane flown. Airplanes covered by FAA definition take all forms, from a Cessna Skylane to an F-51, and it would be foolish to consider *all* high-performance airplanes even remotely similar.

I was introduced to high-performance singles in a unique manner. Back in the early fifties, with some 600 hours, I felt like a real ace, but had only been able to admire the sleek Bonanzas that visited the local airfield. I had never flown one. Then came the annual air tour, preceded by a cocktail and dinner party the evening before takeoff. Our very jovial FAA inspector was there, and he seemed somewhat horrified that a student pilot among us had signed on with every intention of flying his just-purchased Bonanza. The inspector scanned the room, and for some reason I appeared the most likely candidate for instant Bonanza pilot.

He explained to me that if this student pilot flew the airplane on the tour, he'd surely crack it up and hurt himself. My reply was honest: I told him I'd never flown a Bonanza. He responded that it was quite simple and proceeded to tell me how to do it. Young (20) and foolish, I listened intently, decided that it was no hill for a stepper, and agreed to accept the captaincy of the Bonanza.

The first leg the next morning was to an airport 125 miles away. The Bonanza beat my brain to that airport. To begin with, there was no checklist in the airplane. I had to rely on logic, memory, and what little the owner knew about the airplane—which turned out to be virtually nothing. The first revelation came on takeoff. The Bonanza had more torque, or propeller effect, than I had experienced before, and I wondered if one should really have to push that much on the right rudder to keep an airplane on the runway. Raw horsepower—185 total— was the cause of this.

The next revelation came as I started to adjust power for cruise climb. I was looking and not thinking—not looking at the correct gauge—because I held the toggle switch that controlled the prop pitch down too long and the revolutions per minute dropped much below where they should have been. The sound, and a pronounced sag in climb, alerted me to the problem, and it was quickly rectified.

Up at cruise I felt a lot more comfortable in "my" new Bonanza, but I could think only about the landing to come, and with some apprehension. We were headed for a 3,000-foot grass strip; I had been there before and knew that we'd probably be landing downhill. Could such a hot ship be landed in such a small space, downhill?

I didn't realize how aerodynamically clean the Bonanza was until I started the descent for landing. It wanted to go forward much faster than it wanted to go down, and in a flash the airspeed was above the 160 mph top of the green band on the airspeed indicator. Powering back and shallowing the descent I saw that we'd be arriving over the airport with *plenty* of altitude. "Always like to circle the field a couple of times, to look it over," I muttered. In truth, the Bonanza was a lot faster than the Cubs and the Piper Super Cruiser I had been flying, and it simply arrived over the airport before I was ready.

The next reckoning came when it was time to extend the landing gear. I leveled off on the downwind leg, with the power back as far as it would go without blowing the landing gear warning horn. The gear couldn't be extended until the airspeed was back to 125 miles per hour (this was before airplane speed was measured in knots) and the airplane was a long time in slowing to that speed. The resulting pattern was too wide—which was just as well: Further slowing was needed to reach the 105 mph flap speed.

When the FAA inspector had briefed me on how to

fly the Bonanza, he had said that 80 miles per hour would be plenty of speed on final. I had filed that with a grain of salt. It seemed too slow for such a sleek airplane. But with gear and flaps down, the airplane felt very solid at 80. It even floated a little on the landing, but thanks to the delightful longitudinal handling qualities designed into the Bonanza, the touchdown was okay. I was impressed with myself.

The air tour continued and I became more familiar with the airplane. At one point in the tour I had enough sense to decline landing it on a 1,700-foot strip, and I was finally able to make peace with the airplane's speed and aerodynamic cleanliness—and fly it to pattern altitude without circling every airport twice.

Friends later told me that the FAA inspector had said that he would give me my flight instructor rating if I got through the tour without bending the Bonanza. I had been practicing for this rating, and he had previously said he thought I was too young to be an instructor. I never found out whether or not he meant what he said, because he died of a heart attack a few days after the tour. I had to get my CFI from his successor, who was not at all impressed with the Bonanza time (in capital letters) in my logbook. But the experience was valuable, as I learned things about faster airplanes that I'll always remember.

Cockpit Check

Before flying any new airplane, be sure to read through the pilot's operating handbook. This becomes more important as the airplane becomes more complex. Add to it a very thorough cockpit check before attempting a flight. In my Bonanza adventure, I launched as soon as I figured out how to start the engine. Had I sat in the airplane for a while, I surely wouldn't have looked at the wrong instrument while changing the prop pitch. And I'd have had a better handle on airspeed limits than I had by looking at the dial at the eleventh hour.

The cockpit check is important, because there's not an overabundance of standardization in light airplanes. For example, on new Bonanzas the landing gear switch is to the right of the power controls and the flap switch to the left. The reverse is true on Cessnas (and on some Beech twins). The Bonanza's throttle and prop controls are next to each other, and the mixture control is below the throttle. Cessna lines them up, and Piper has a "twin style" power console. Some throttles are push-pull, others vernier. On some airplanes, the instruments that monitor power—the manifold pressure gauge and tachometer—are left to right in the same order as the knobs that control them (throttle = manifold pressure, prop control = rpm); on others they are in reverse order. Some airplanes have cowl flaps and some don't. If you first study the handbook and then spend a few minutes finding things in the cockpit, flying will be easier. If the person checking you out in an airplane tends to rush through this, tell him to slow down.

The Purloined Mike

One day I left the Cessna plant to ferry a new 310 back to the East Coast. I wasn't given a checkout in the airplane because I was considered current, having flown one a few months before. But I treated it like a new-to-me airplane. I read the book the night before and spent a few minutes in the airplane becoming reacquainted with the various switches and instruments—*before* even reaching for the checklist.

I read through the list once before touching anything. I felt ready, and then went back to the beginning to bring the airplane to life by the book. Everything went perfectly; the engines started after only a few blades, the radio was set, everything was double checked, and I was ready to call for taxi clearance. Ready to fly away home.

Hmmmm, the microphone? Look? I searched all over the cockpit. I couldn't find it, and with my face as red as after a weekend at the beach, I shut the airplane down and went to the office to tell them that there was no microphone in the airplane.

"Oh, yes, there is. On that model it's mounted at the right front edge of the pilot's seat."

That information was given in the book, but I must have sailed right by it, thinking that any fool knows where to find the trusty mike.

Retractables

Moving up to a retractable-gear airplane is the usual first step up to a high-performance airplane, and there's a lot more to consider than simply putting the wheels down before landing. One of the first questions, in fact, is when to retract the landing gear after takeoff.

Some pilots have been known to retract the gear immediately after liftoff. More than one such pilot has had his airplane suffer a sinking spell after the gear is up —or after it has started up—and has bent the propeller tips and more as a result. No question, the landing gear *should not* be retracted at the instant of liftoff. The *only* benefit that might accrue would be to the pilot's ego— how sharp it looked to get the gear up immediately.

It is often recommended that the landing gear shouldn't be retracted until a landing is no longer possible straight ahead on the remaining runway. This is the best procedure in many cases, but it is difficult to tell precisely when such a point is reached. Some airplanes simply do not climb well with the gear down, and if departure is from a long runway, you might drag that gear through a lot of air while wondering when to retract.

Options are always good to have in flying. Why not retract the gear when high enough to have time to extend it again in case you have to get back down quickly? That way, both options—gear down and gear up—would pertain. The latter might be preferable only if minimum stopping distance after touchdown is critical or if an off-airport site looked best for a gear-up landing.

An oft-told story on quick gear retraction is worth

repeating: A pilot took off on a long runway, and soon after liftoff an improperly secured cabin door popped open. This always makes a lot of noise, shakes the tail, and is generally disconcerting. There was plenty of runway ahead, so the pilot elected to land straight ahead and close the door. The only trouble was that he had already retracted the landing gear. It was an expensive oversight.

Retractables, Clean and Dirty

En route, a retractable may not seem any different from other airplanes, but there's a situation in which its aerodynamic cleanliness can be a hazard.

When a pilot with little or no instrument training (or one with a rusty instrument rating) enters cloud, the result is sometimes loss of control. A lateral upset generally precedes (and causes) a longitudinal upset. Once the airplane is out of control in a spiral, gravity holds sway and the airplane heads downward at a high rate of speed and descent. Speed builds rapidly, and a retractable can be over the redline in a very short time. The outcome is often a structural failure. The same scenario often applies when an experienced pilot enters extremely turbulent conditions, as might be found in the vicinity of a thunderstorm, and loses control of the airplane. Any loss of control in an aerodynamically clean airplane is serious, and the options and odds for survival decrease rapidly as the speed increases.

Then why not extend the landing gear *before* things get out of hand? The airplane is no longer aerodynamically clean and slippery, and options and odds are better.

A classic case of the use of the landing gear in time of distress was found with the 727 that suffered an upset at 39,000 feet and descended out of control to below 10,000 feet. The crew lost control, and the airplane was in a screaming dive for earth. Flaps and spoilers were used in trying to regain control of the big airplane, but they didn't get things back in hand until the landing gear was extended. Apparently this provided the extra measure of drag necessary to stabilize things and permit a recovery from the unusual attitude.

Beech installs a placard on the Bonanza sun visor giving good advice for a pilot who finds himself on top with no VFR way down through. The word is to tell the FAA what's happening, extend the gear and flaps, trim the airplane for a normal approach speed, establish a descent, and *keep the wings level with the rudder pedals* as the airplane descends through the clouds. That's simple, and it can be a real lifesaver.

Good descent planning requires common sense and recognition that the retractable you are flying is faster than the fixed-gear airplane last flown. Descents are simple mathematical exercises, and a pilot who lets an airplane get ahead of him in a descent has simply forgotten to do a little simple arithmetic. Landing a retractable is like landing anything else—if the wheels are down.

Handling High-Performance "Feel"

Handling qualities are important as we move into higher-performance airplanes, because there *are* differences. As the speed and size of the airplane increases, engineers find themselves resorting to such things as downsprings and bobweights to improve longitudinal stability. While these additions are certainly helpful, you do notice them when flying.

Consider the Skyhawk and Skylane, two airplanes that look a lot alike. The Skylane qualifies as a high-performance airplane. It is bigger and heavier and more powerful. And while the Skyhawk has very good longitudinal stability without devices, the Skylane has a downspring. To pull the wheel back, you have to pull against this spring. It increases the force required to pull Gs, or to move the airplane away from the trim airspeed, thus making the airplane more stable longitudinally.

The spring is most evident on landing, especially on landing with the center of gravity forward—when the airplane is normally rather strong longitudinally. If the airplane is trimmed for a 68-knot glide, the force required to flare will be greater than required for a Skyhawk, and probably for most any other airplane that doesn't fit into the high-performance category. You'll be pulling against the spring as well as the forces on the elevator. Also, the feel of an airplane improperly trimmed for the approach will be more pronounced in an airplane with a downspring. You'll have to pull against the out-of-trim condition *and* against the spring. The result is that pilots sometimes don't use enough back pressure when landing an airplane like a Skylane and let it touch down in a level attitude. That's hard on nosewheels.

The Cardinal RG is longitudinally interesting. The airplane has a very big stabilator, and it must have been challenging for the engineers to work out optimum control forces. If a very "light" stabilator had been chosen

—that is, a little effort on the wheel would move the big stabilator quite a distance—pilots would have tended to overcontrol longitudinally. At speeds above maneuvering speed, it would have been easy to overstress the airframe. So they made the stabilator control rather heavy, or stiff. This gives the airplane nice handling qualities en route, and while the landing qualities of the airplane are adequate, there are enough differences between this and other airplanes to cause some consternation.

When I bought my Cardinal RG, my first few landings were terrible. I decided that if I didn't spend a little time learning to land the airplane, I might break it in two during the first 100 hours. It took only a few practice landings to learn to do it.

In letting other pilots fly the airplane, I saw only one make an acceptable landing on the first try. John Olcott, engineer and editor as well as experimental test pilot, landed it perfectly. He says it was luck, but I say it was in being able to anticipate responses based on what he felt during the flight at altitude.

The Cardinal RG gives a first impression of being stiff longitudinally. Then, when it's time to flare, the pilot expects a strong pull to flare and land the airplane. But the stabilator is so big and powerful that it needs to move only a little to greatly change the pitch of the airplane. So there can be excessive changes in pitch attitude at flare. The airplane balloons and the pilot's reaction is to pull back with less force as the airplane settles back toward the runway. By this time the airplane is slower, and the stabilator a little less effective. *Bang,* it hits hard and level.

What the Cardinal RG needs is a fine eye for pitch

change in relation to control force applied and a pilot aware of the airplane's idiosyncracies. Even awareness isn't a cure-all, as I found when renting a Cardinal RG for a trip a few months after I sold mine.

When I started discussing the rental with the man behind the counter, he was impressed with the fact that I had over 1,000 hours in a Cardinal RG. His tone was even apologetic as he explained that their insurance requirements necessitated a checkout in the airplane.

The flight instructor assigned the duty mentioned that I probably had more Cardinal RG time than he had total time. In reply, I went through the little speech about the hours in the log not counting for nearly as much as the next hour. Then I proved the point by making a truly lousy landing in the Cardinal. I had forgotten. In fact, on the rental trip I made a total of three more landings in the airplane and never did get a really good one.

T-tails

Other interesting airplanes in this regard are the T-tail Lance and the T-tail Arrow—especially the Lance. A pilot who has to pull hard to fly a Cardinal RG adequately could find flying the original T-tail Lance an elusive proposition. Why? Because the stabilator atop that Lance's T-tail is somewhat undersize, and the Lance has a greater center-of-gravity range. This changes the rules.

On takeoff many pilots notice that the Lance seems difficult to get off the ground, especially with forward loading. They pull back on the control wheel at what seems the proper speed, yet nothing happens. Then, a moment later, the airplane pitches up and abruptly flies into the air. Some porpoising might follow as the pilot overcontrols to get the airplane in proper pitch attitude.

There is logic to all this. At the start, everything's just fine as the Lance rolls along the runway building speed. But when the pilot decides to fly and pulls back on the wheel, he gets no reaction and pulls back some more; this obviously stalls the stabilator. Remember, it is small in relation to the total airplane. If you tilt it at a big angle to the relative wind, a stabilator will stall just as a wing does. The stalled stabilator might have little initial effect on the pitch attitude of the airplane, but as the airplane gets going faster, the stabilator starts working. The pitch attitude changes slightly, nose up. As the pitch changes, it reduces the angle-of-attack of the stabilator. Reduced enough, the stabilator recovers and flies out of its stall and becomes fully effective. What the pilot considered a sluggish, ineffective control comes alive. The sensitivity of it changes markedly, and overcontrolling and porpoising are natural consequences as the pilot tries to cope with the "new" feel of the stabilator.

A smooth (but mechanical) takeoff in a T-tail Lance is when the pilot reaches a certain speed, deflects the stabilator a certain amount, and simply waits for the airplane to fly off the ground.

The T-tail Arrow IV exhibits some of these takeoff tendencies, but they are minimized by an inverted slot in the stabilator. The inverted slot retards the stall of at least a portion of the stabilator and makes stabilator effectiveness develop more evenly as the airplane accelerates.

T-tail Lance landings are easier to handle. It is the difference between starting at zero and accelerating to a speed where the stabilator suddenly becomes effective, as on a takeoff, and dealing with slowly diminishing effectiveness as on a landing. On takeoff, you don't do much with the stabilator until ready to fly. On landing, you had better use it continuously all the way in.

Loading and Performance

I used to own a Cherokee Six, a good example of a relatively simple airplane that requires more than a basic understanding. This is true of all the six- and seven-place singles because of the wide cg loading and ranges.

The Cherokee Six with two or three people on board is a lot like any Cherokee. Load it to gross, though, and it grows a different set of stripes. At aft cg, the airplane is tender longitudinally. The elevator forces are very light. And while the takeoff and climb performance is excellent at light weight, it is somewhat lethargic when loaded. The wide range in weight means that the takeoff, climb, maneuvering, and approach speeds need careful adjustment. If the gross weight approach speed is used at the end of a long solo hop, the airplane will float halfway across the airport. The rule of thumb for reducing performance speeds—half the percentage of weight below gross—worked well for me. For example, if 20 percent below gross, I'd shave all speeds by 10 percent.

In any airplane with a big useful load, it's important to fly at gross weight during a checkout. That eliminates any surprises later.

Looks Don't Deceive

When checking out in a high-performance airplane, spend some time just looking at the airplane from outside. Its basic appearance can provide a lot of clues to the way an airplane is going to handle.

If the nosewheel of the airplane is far forward and the main gear far back, for a long wheelbase, the forces required to lift off might be rather high. Main gear far back usually means that instead of rotating the airplane around a point close to the center of lift you'll be heaving the nose off the ground.

The wing's angle in relation to the ground is another clue. If the airplane rests on the ground with the wing at a positive angle-of-attack, the wing will start developing lift almost as soon as the airplane starts moving forward. The airplane will quickly become progressively lighter on the wheels as it accelerates. Nothing wrong with this, though it does require compensation in crosswinds. As the airplane gets lighter on the wheels, it will be more likely to skitter sideways a little, and proper aileron use becomes important. An airplane that sits and runs at a positive angle is more likely to fly itself off and require very little back pressure for liftoff.

When an airplane rests with the wing at a neutral or negative angle-of-attack, then lift development won't be as noticeable as the airplane accelerates. And it is likely to require more force to rotate the airplane to a flying attitude.

If an airplane has small elevator surfaces or a small stabilator in relation to the airplane size, then pitch control is likely to be different, as on the T-tail Lance. If an airplane has very large tail surfaces, response usually is more conventional. If the fuselage appears long, it'll probably ride well in rough air. If the fuselage is short, it'll wiggle with the bumps.

Systems

High-performance singles don't have highly complex systems, but each airplane's systems usually have features that need to be understood. A retractable's emergency gear extension system should be tried at least once in practice. Fuel pumps are worth study, too. On some airplanes the fuel pump is used for takeoff and landing, while on others it is used only for starting or when the engine-driven pump fails.

Today's high-performance single has the performance capability of yesterday's airliner, and too many general-aviation pilots take this lightly. While you can fly a basic single—like a Skyhawk or Cherokee—with basic flying skills, the high-performance airplane demands a procedural approach. It isn't a toy, and it isn't designed to be "fun" to fly. Once past the basic light single, airplanes are designed to go as fast as possible, fly as far as possible, and carry as big a load as possible. All this is available, at a good level of safety, *only* if the airplane is treated as the serious machine it really is.

b
Flying the Twins

An airplane with two engines seems to make a promise. It appears to have redundant powerplants, as if to suggest that it will fly very well on either one. That promise is kept by airline and business jet aircraft, which by regulation must be operated in such a way that the flight can be completed successfully on one engine. Up to a predetermined point on the takeoff run, the captain will abort in case of an engine problem with adequate runway remaining in which to stop. Once past a speed called V_1, or decision speed, the airplane has the capability of successfully continuing the takeoff despite an engine failure. In the course of preflight planning, pilots are required to compute whether or not the runway is long enough for this procedure. If it comes up short, the weight of the airplane must be reduced accordingly.

The first general aviation twins did not come close to such performance. One, the Cessna T-50 (or UC-78, or "Bamboo Bomber"), could not even maintain altitude on one engine.

I got my multiengine rating in a T-50 in the early fifties. The airplane was nice to fly and supremely forgiving, but its engine-out lethargy was plain to see. In practicing for the check ride, the engine-out drill consisted of pulling one throttle back and marveling that the airplane didn't immediately plummet to the ground. But it was all forward and down.

On my check ride, the T-50 was trundling down a rather short runway—about halfway to the speed at which it would fly—when the examiner pulled one throttle back and asked me what I was going to do. Only one thought passed through my mind, and I took fast action while saying "If I don't push that throttle back in, we'll crash." And we surely would have. The brakes on the airplane were not very good, there wasn't much runway left ahead, and flying on one engine was not one of the options.

A friend experienced the T-50's engine-out performance firsthand one day when an oil seal blew out of one engine and reduced it to a cold and greasy hunk of metal. He was 50 miles into a 120-mile trip and was over the pine forests of southern Arkansas. It was before federal aid for airports, and the only suitable fields were the departure point and the destination. Returning to the departure point seemed the best bet, and he headed there with about 3,000 feet of altitude.

The pilot wasn't much closer to the departure airport when he could see that the peanut butter and the bread would not come out even. In covering 10 percent of the distance, the airplane had given up 20 percent of the altitude.

A very short strip near a small town along the way was the only hope. He had used it once before, and while it was hardly more than a break in the forest, he believed he could get the airplane on the ground and *then worry* about stopping it.

As the little strip came into view, he saw that it was going to be close. The gear and flaps were extended at the last minute, and he reduced power only slightly on the good engine before touchdown. The stop was successful, though the tires did tear up the turf during his maximum braking. The sheriff came quickly after hearing that a big "transport plane" had landed at the local airport, and made his official observation to the local paper: "Why, our little strip was so short that the pilot had to shut off one engine to make the plane fit."

Fortunately, new twins do better than the old T-50. But there are still enough compromises for any statement that a twin has redundant power to be taken with a grain of salt.

The FAA regulations covering engine-out performance for twins are minimal. If a twin has a gross weight of 6,000 pounds or less and a stalling speed of 61 knots or less, there is no requirement that it climb on one engine at any weight or density altitude. If the stalling speed of a piston-engine twin is more than 61 knots, the airplane must—in standard air at 5,000 feet—be able to maintain a steady climb of .027 times the square of the stalling speed with the gear up, flaps in the most advantageous position, and everything else set to maximum advantage. (For example, if the airplane stalls at 70 knots, square that to 4,900 and multiply by .027 for a required climb rate of 132.3 feet per minute at 5,000 feet.) There is no requirement for complete engine-out performance data on light twins. All that's required is an engine-out rate of climb (or descent) figure.

The climb requirement at 5,000 feet for airplanes that stall at more than 61 knots is one reason why so many twins have turbocharged engines. There's no way to get the gross weight up to a practical value and climb as required at 5,000 feet without having full power available at that altitude. And when you do find a nonturbocharged twin with a stalling speed of over 61 knots, the single-engine climb at sea level is pretty good. The Cessna 310, for example, climbs 340 feet per minute, the Piper Aerostar 340, and the Beech Baron climbs 397 feet per minute on one engine at sea level: all at gross weight in standard conditions. Turbocharged twins don't have to do so well at sea level to meet the 5,000-foot requirement, and most are therefore pretty soft. The Aerostar drops back to 254 feet per minute in the turbocharged version, and the Piper Navajo Chieftain is 230. The Cessna turbocharged 310 remains in very good stead, at 390 feet per minute, but this is a result of using the same gross weight as the nonturbocharged airplane.

Flying from a sea level airport in a twin with a stalling speed above 61 knots, you are better off with one that isn't turbocharged. But go to higher density altitudes and turbocharging is the ticket.

Any way you slice it, it is clear that airplane designers place more emphasis on a payload as high as possible than on single-engine performance above the minimum required by FAA regulations. This is okay: It gives the pilot the option of flying the airplane at lighter weights for better engine-out performance.

The 1979 production light twins with stalling speeds low enough to be exempt from FAA climb requirements were the Cougar, the Seminole, the Duchess, and the venerable old Aztec. The Cougar's engine-out rate of climb is 200 feet per minute, the Seminole's 217 feet per minute, and the Duchess and Aztec will go up 235 feet per minute on one engine at gross at sea level in standard conditions. Apparently, the FAA feels that if a twin's stalling speed is as low as a single's (which also must be 61 knots or less), then engine-out performance isn't critical for the twin. But a single with the twin's engine-out rate of climb would be neither licensable nor marketable.

Regardless of what kind of airplane you fly, single or twin, it is most important to know beforehand what the airplane can and cannot do. Trying to make an airplane do something beyond its capability is hazardous, and the outrageous engine-failure-related accident record in general aviation multiengine flying is proof aplenty that many pilots do not understand what their airplanes cannot do. After engine failure they persist in trying to extract performance that is simply not there. The usual outcome is a loss of control.

Unless the twin's handbook contains data for determining the distance required to accelerate to takeoff decision speed and *stop, as well as* data showing the distance required to accelerate to that speed, have one engine fail, and *then* continue the takeoff up and over an obstacle, then the pilot has no way of determining what the airplane will or will not do if an engine fails during takeoff.

There is simply no way for the pilot to *know* what to do if an engine fails unless this information is available to him.

The Accelerate/Stop Myth

For years, most multiengine airplane handbooks have contained accelerate/stop distance data offered as a measure of performance. This means very little, because in comparing the airplane's performance with the runway length, you must calculate both accelerate/stop and accelerate/go distances and consider the longer of the two as the limiting factor. Also, many accelerate/stop charts are even less meaningful, because they cover acceleration to a speed well *below* the speed on which engine-out climb performance is based. For example, the chart might show the distance required to accelerate an airplane to 80 knots and stop even though the airplane has an engine-out best angle-of-climb speed of 90 knots. That 10-knot difference is the zone of total mystery. Simply stated, an accelerate/stop chart means nothing unless you have an accelerate/go chart to go with it.

A misused speed number on twins is V_{mc}, or minimum control speed. By definition this is the minimum speed at which the airplane is controllable with the "critical" engine (the left one on twins with a pair of right rotating engines) inoperative. V_{mc} is as much a measure of rudder effectiveness as anything else, and it is meaningful *only* in relation to control. It has nothing to do with performance and has nothing to do with any decision to continue a takeoff after an engine failure. Certainly if one engine fails when the airspeed is below V_{mc}, the takeoff must be aborted; the only way to make the airplane go straight is with a reduction of power on the operative engine. This is made clear by the airplane: it won't go straight, and it will convince you of that fact quickly. It is not like performance, which needs to be calculated in advance to know what to expect if you decide to continue the takeoff.

Planning for Performance Capability

Preflight planning for every takeoff in a multiengine airplane should include a calculation of the airplane's performance capability in relation to runway length. If this is not done, you won't have any workable idea of what to do should engine trouble develop.

For example, let's take the turbocharged Cessna 340A. The handbook on this airplane is very comprehensive, and there's no excuse for a 340A pilot to be ignorant of his options on takeoff.

At gross weight, the decision speed on a 340A is 91 knots. If the temperature is +20 C, the airplane will accelerate to that speed and then stop in a total distance of 2,990 feet at sea level. Moving to the next page in the handbook, we see that the airplane will accelerate to 91 knots, suffer an engine failure at that point, and then continue to an altitude of 50 feet in a total distance of 4,360 feet at sea level. That's based on engines at full power before releasing the brakes, prompt retraction of the landing gear, and prompt feathering of the offending propeller. So, under the given conditions, the 340A should not be operated off a runway shorter than 4,360 feet—unless the pilot is willing to accept the inevitability of a crash if either engine fails at a critical time in the takeoff. If the runway is longer than the accelerate/stop distance but shorter than accelerate/go, the critical time span lasts from the moment 91 knots is reached until the airplane is off, cleaned up, at an altitude above all obstructions, and at or above the 100-knot best single-engine rate-of-climb speed.

If the pilot is flying from a 3,000-foot runway and wants accelerate/go capability, the performance chart shows that the weight of the airplane must be limited to about 5,500 pounds. (Maximum takeoff weight is 5,990 pounds.) This is certainly the best option.

The air transportation department at Cessna requires that their pilots prepare a takeoff data card for every flight. This shows the pertinent speeds as well as the required runway lengths. It is good practice and should be used by all pilots flying twins.

Let's also consider the Beech Duchess, which has a 71-knot decision speed. At +20 C at sea level, the Duchess will accelerate to 71 and stop in 2,500 feet. It will accelerate to 71, lose an engine, and then go on over the obstacle in about 5,000 feet. See why accelerate/stop doesn't mean much? The only time it would be down close to accelerate/go on the Duchess would be on a cold day, 800 pounds below gross, with a 20-knot headwind. And the Duchess offers good engine-out performance when compared with the current light twins stalling at 61 knots or less.

Engine-out Climb Gradient

Once calculations have been made for accelerate/-stop and accelerate/go, engine-out climb gradient is next. At +20 C, sea level and gross weight, our 340A is going to climb 295 feet per minute at 100 knots. There's going to be a flat spot in climb because the airplane, after clearing the obstacle, must accelerate from 91 to 100 knots. Once at 100, and disregarding any wind effect on groundspeed, this performance means that the airplane will climb 177 feet for each nautical mile traveled. (100 knots is 1.666 miles per minute; divide that into 295 for the gradient in feet per mile.) Feet per mile is the only way to evaluate engine-out climb, because it's what tells you how high you'll be where; this is the critical question when clearing obstructions.

Under the same conditions, the Duchess will climb 210 feet per minute, but its climb speed is lower: 85 knots. The climb gradient is 148 feet per nautical mile.

For an example of the rewards of a turbocharged twin, consider both airplanes at 5,000 feet and +20 C. The Duchess gradient is down to 42 feet per nautical mile while the 340A's is still 96 feet per mile.

None of these figures is very impressive. If, for example, you take off at sea level, lose an engine at 200 feet, and there's a 1,000-foot-above-ground-level obstruction that needs to be cleared, you'll need 5.6 flying miles to clear it in the 340A and 6.7 flying miles to clear it in the Duchess. If it is IFR, and if, after trouble on a departure, you fly the instrument approach procedure backwards (fly it *away* from the airport instead of toward the airport to get the best possible deal on obstruction) you still won't have a very good deal. Descent gradients on in-strument approaches can run as high as 400 feet per mile from minimum safe altitudes to the airport. A common glideslope descent on an ILS is over 300 feet per nautical mile. Light twins just won't match those numbers on one engine.

When operating with an engine out, it's obvious that the path of lowest resistance must be followed.

The plan for a multiengine takeoff begins to take form as we complete our calculations. If we are to have performance capability in the event of engine failure (which is the purpose of flying a twin in the first place), the runway length should equal the longer of the ac-celerate/stop and accelerate/go distances. The speeds to consider are decision speed, speed over the 50-foot obstacle, and the best single-engine rate-of-climb speed. V_{mc} doesn't figure in, as it is below decision speed and an engine failure below that number is a mandate to abort the takeoff.

Planning Pays Off

It's smart to integrate your numbers into an overall plan for normal takeoff—and review the plan just before takeoff.

Hold the brakes, full power, release the brakes, and be ready to cut power and stop if anything appears amiss before reaching decision speed. Start to lift off at deci-sion speed. Retract the landing gear when a positive rate of climb is evident, climb at the obstacle clearance speed until above all obstacles, then accelerate to the best rate-of-climb speed for operation on both engines. Fly that speed at full power until at a comfortable altitude (at least 1,000 feet above the ground), then move on to cruise climb.

There is a good alternate technique followed by a lot of experienced multiengine pilots. Let the landing gear —not decision speed—be the determinant of stop or go in case of engine trouble on takeoff. If there's a power loss before the gear switch is put in the up position, then the takeoff is aborted. Put it back on the ground and stop.

This takes one step out of the emergency procedures for configuring the airplane for engine-out operation. You don't have to put the gear up, because it is already on the way up if the decision is made to continue flying. It is true that using gear retraction as the decision point means the airplane won't really stop within the distance calculated for acceleration/stop. But entry into the rough should be at a slow speed, which helps—unless the airport is like the one at Charleston, West Virginia, with a precipitous drop off the end of each runway.

The success of this procedure is predicated on an airplane operated within its accelerate/go envelope. If the pilot unwisely decides to fly from a runway shorter than the accelerate/go distance and then decides to continue the takeoff after the gear switch is put in the up position, the airplane might impact the old oak tree only a few feet up its trunk.

Don't Fool with the Numbers

Let's pause to consider the philosophy of operating twins only from runways equal to the greater of the accelerate/stop and accelerate/go distances.

Safety is the key word. Operated this way the twin has a better safety potential than any single. Throw this advantage to the wind, and the single-engine airplane has more takeoff safety potential, simply because it has half the chance of engine failure. Also, the stalling speed of the single is usually lower, and it will contact at a slower speed in any forced landing. There's usually less mass to decelerate in the single, too. And finally, the record shows that there's a strong possibility of loss of control when a pilot attempts to continue engine-out flight in a twin when performance capability isn't up to the task. The pilot might tend to leave the operative engine running at full power while trying to "make" the airplane fly over the obstructions. Slower and slower it goes until directional control starts to deteriorate. Here the pilot faces the difficult task of reducing angle-of-attack and power at the same time. It's not difficult to do this in the simulated atmosphere in which we are trained, but it is *very* difficult to do when trees are looming large. The illusion of redundant power is seductive, and to the last it seems that the airplane will keep flying. It's a situation that has only bad, worse, and worst outcomes, and the only solution is to fly from runways equal to the accelerate/stop or accelerate/go limit, whichever is greater.

What if detailed performance data for the twin you are about to fly are not available? First, I would try my best to get the information from the manufacturer. Failing that, I would find a complete operating handbook for an airplane of similar weight, wing loading, power loading, and general performance, and use it to develop rough guidelines for the airplane's performance. That's about the best you can do. If, with this information in hand, you decide to take off from runways that do not afford the engine-out advantage, do so while accepting the risk involved. With singles, that risk is obvious—cut and dried. Twins are a different matter. The risk for each takeoff must be calculated.

Engine-out Technique

Back to our plan for takeoff. If an engine fails before 91 knots in our 340A, it will be obvious and you'll quickly have to pull the throttles back, work hard at directional control, and get on the brakes.

Failure *after* liftoff is not so simple. The gear is up or coming up, and if the decision is made to keep flying, feathering the propeller becomes the project of the moment. Which one to feather? "Dead foot, dead engine" has been the battle cry for years, and it has withstood the test of time. If you are pushing with your left foot—and you'll surely be pushing hard if the airplane is going straight—the right engine is the dead one. Your right foot is not doing anything. You are now sure it *is* the right engine. But don't grab the prop control and snap it back to feather without double checking. The throttle is the thing to use next. Pull the right throttle back, smoothly and not too rapidly. If nothing happens when you pull it back, you have confirmed that the right engine is inoperative. Then pull the right mixture back and, finally, feather the propeller on the right engine. (There's a checklist in the handbook to complete the securing of the engine. All these additional steps should be taken, but probably not until the obstructions are cleared. It's time to fly.) If the book says that 91 knots should be maintained until obstacles are cleared, maintain 91 knots. Not 90 or 92, but 91. Once over the obstructions, accelerate to the best rate-of-climb speed for single-engine operation (the blue line) and fly that until at enough altitude to return safely and land. As we saw in our discussion of climb gradient, this will take a while.

How high should the airplane be before you start back to the airport? That depends on obstructions, how fast the airplane is climbing, wind, and turbulence. Only one thing is certain: When a turn is started, the rate of climb will suffer. A turn should not be started until you are ready to fly level or accept a gentle rate of descent. And the turn should not be steep. Again using the 340A as an example, flying wings-level at 100 knots there's a 17-knot buffer between indicated and stalling speed. Bank 40 degrees and that margin drops to six knots. A bit of turbulence could cause the airplane to stall with only a six-knot margin.

In flight-testing a light twin, I've loaded one to gross, failed an engine at liftoff, feathered the prop, and continued the takeoff. Believe me, it is a very delicate maneuver, one in which you must do everything correctly. I learned that most things are obvious. If the airplane is climbing, you can see and feel it. If it is going to make it over the obstacle, it is apparent. If it isn't doing well, that is apparent, too.

What if the decision is made to go, and then the airplane fails to perform as you thought it would? This is clearly a time to admit defeat and start cutting losses. Where is the best possible landing area? Find it and head for it. This is when you'd be better off in a twin like the Duchess, with a lower approach and stalling speed.

I was going through the engine-out business in a Cougar one day and had an interesting session with the options. One throttle was retarded to zero-thrust, a setting that is supposed to simulate a feathered prop. The airplane was about 100 feet above the trees and was just barely climbing. In fact, it would climb a little and then settle a little. There was some net gain, but it was almost imperceptible. Had the situation been actual, I would have landed the airplane in any big field I passed over.

Why do that instead of nursing it back to the runway? For one thing, I was flying into the wind and could land straight ahead, into the wind. Had I come around on a downwind leg and the airplane wasn't able to clear the trees and terrain, I certainly wouldn't have been able to make another 180 back into the wind. And when depositing an airplane in impromptu surroundings, I'd rather do it *into the wind.*

In another instance and airplane, I departed the runway, feathered a prop right after liftoff, and after examining the situation, started a turn back to the airport when only a mile away. I had taken off to the southeast; I came back and landed to the north. The wind that day was calm and there was no turbulence. Again, I was testing a new airplane. Things like that shouldn't be done without careful planning, unfeathering accumulators to get the engine back on line quickly, big airports with plenty of open spaces around, and pilots who are very familiar with the airplane. We'll discuss practicing in more detail in the next chapter.

In actual engine cuts, there is something in favor of the twin—at least momentarily. Once off, there is upward and forward momentum. When an engine is cut, the airplane doesn't lose this upward and forward momentum immediately. It might even allow enough time to identify the dead engine and feather its propeller. That helps, but it could also deceive the pilot. If things looked good for a moment after engine failure, it might cause him to opt for continuation; the sag after the momentum could prove disastrous.

Regardless of preflight calculations, the airspeed indicator must be used as the absolute determinant of the pitch attitude of the airplane—especially if the airspeed is below critical speed—91 knots in our 340A example— or if the airspeed is decaying. In either case, the angle-of-attack of the airplane must be reduced. It is that simple. If the airspeed is enough below 91 knots to cause directional control problems, power can be reduced on the good engine to get that in hand. But this power reduction must be done *as* angle-of-attack is reduced. Keeping the airspeed above the stall (and above minimum control speed) is the only thing that will enable the airplane to be landed under control.

Partial Failure on Takeoff

When trouble comes after liftoff but at a critical time, what if the power problem isn't clear-cut? When an engine is surging, it is difficult to tell immediately which one it is. Or maybe one engine suffered a *partial* power loss, as a turbocharged engine might do if a turbo hose comes loose.

In this situation I would use rudder pressure to make the airplane fly straight and I would scan the airspeed and vertical speed indications to decide whether or not to secure an engine. If at or close to obstacle clearance speed and with much less than full rudder required to go straight and a healthy climb shown on the vertical speed, I would leave things alone and climb for a while. Once above all obstacles, I might feather the engine if it was still cutting up. Or given the chance for an immediate return and landing, I might let it contribute whatever it could. The only penalty in *not* feathering might come if the problem degenerated into something that would prevent feathering. On some airplanes a complete loss of oil pressure precludes feathering, and you'd feather early if you saw a large quantity of oil streaming back from a rattling powerplant.

Many engine failures develop not from mechanical situations but from pilot-induced problems, such as incorrect fuel management. Thus when an engine fails it's usually suggested that you investigate things that might have caused the failure. Change the setting of the fuel pump switches, check the mags, and fiddle with the mixture. All this takes time—more time than would be available after a failure at a critical moment—and you had best do the preflight and the pretakeoff checklist thoroughly enough to have confidence that everything is correct for takeoff. Then you'll have eliminated "pilot error" as the cause of any trouble.

Timing is critical on takeoff. Success or failure is based on what happens in just a few seconds. Consider the tale of the fellow who had an engine failure on takeoff

in a single—because there was no fuel in the tank selected. The gauge was stuck and read "full," and he
didn't suspect the cause. His instinctive reaction was to
head the airplane for an open space and land it. *If* he had
switched tanks, the forced landing would not have been
necessary, but this was a split-second event. The mind
can handle just so much at once, and I'll always bet on
the pilot who flies the airplane first and handles the other
things as time allows. Maybe you could say that it was
pilot error for this lad not to switch tanks, but he is as
healthy as ever today (though the airplane is not); if he
had spent his three seconds (or whatever) switching
tanks, and the engine hadn't caught quickly, he might
have wound up in an apartment house instead of an open
area.

Rudder Versus Yaw

Remember that the product in a twin of one engine
going and the other dead is *yaw*. The control to use in
countering yaw is the rudder. Use it to the fullest. If a lot
of aileron is being used to keep an engine-out twin going
straight, replace it with rudder. If there is not enough
rudder available to keep it straight, then the airplane is
flying too slowly. Reduce angle-of-attack.

One relatively minor thing that can be done to improve climb and help control is to bank slightly toward
the operating engine—about five degrees is enough. If it
takes a lot of aileron to maintain this slight bank, go back
to the rudder to control yaw. Use more rudder and less
aileron.

The business about yaw emphasizes a basic of staying

out of trouble in a twin. One of the causes of a spin is yawing moment at the stall. You don't want to get close to the stall in a twin with an engine out: As directional control is lost, the ingredients for a spin will always be there—waiting.

Engine Failure En Route

An engine failure en route is no immediate big deal in a twin. Demonstrator pilots often show how you can hold 'er straight with ailerons while feathering and while watching the speed settle back to a reasonable engine-out cruise. This is all well and good. It gives you time to prepare for landing, and an engine-out landing in a twin is as challenging and difficult as a dead-stick landing in a single. There's a little more energy to manage in the twin, but the consequences of mismanagement can be just as serious.

I once had an engine failure in an old Twin Bonanza (it was a new airplane then)—one of those easy en route ones. The engine ran rough and started trailing smoke through the oil breather on top of the cowling. It also started rattling more than usual, and I decided to secure the engine.

I was halfway into a long flight and wanted the closest place with a runway of reasonable length. This turned out to be Charleston, West Virginia, an airport that I had never used before that day. I was IFR, and the controller quickly gave me a clearance to Charleston. I told him I'd like to maintain my 8,000-foot cruising level as long as possible. That was approved.

The ceiling was three or four thousand feet, and I broke out a few miles from the airport. Never having been there before, I didn't know that the Charleston airport had been formed by taking the top off a mountain and that there was a drop at the end of each runway. It was a situation demanding top performance, and I was glad that I had spent a lot of time practicing engine-out landings in the Twin Bonanza. It all worked out okay, but in retrospect I wish I had been a little more knowledgeable about the airport. I wouldn't have gone on to the original destination, but I think I would have flown a few extra miles to find an airport that offered more margin for error on an engine-out approach.

Approach and Landing

There are two schools of thought on how to make an engine-out approach and landing. One says that the energy stored in altitude should be hoarded and given up reluctantly—meaning a higher and steeper-than-normal approach. The other advises a normal approach with the only possible exception being a somewhat later point in the pattern for gear and flap extension. I'd opt for the latter method; I feel that an engine-out landing can and should be the same as a normal landing—if normal landings are done properly.

Every approach in a twin should be flown as if it is an engine-out approach. Then if the day ever comes, you'll know what you're doing. For example, when flying normal multiengine approaches in the 340A I'd use a power setting of 2,450 rpm and 20 inches from downwind leg on. This is less than 50 percent power—*less* than the power that would be available if the airplane were flown on one engine. Plan everything so that this amount of power is not exceeded on a normal approach. Any time you drop the gear and flaps too early and find that this amount of power has to be exceeded in the approach, you'll know that the approach would not have been made with an engine out. It's a simple way of making *every* approach count as engine-out practice.

On downwind I like to have the gear extended when opposite the landing point. This will start the airplane down. I fly a normal base, a normally steep approach, and when on what appears to be the desired approach slope, I start extending the flaps. Once the flaps are past approach setting, I consider myself committed to the landing. Going below 1,000 feet above the ground might also be a commitment to land. That momentum we talked about on the takeoff is again there, but on approach it is in a downward direction. A go-around on one engine is a delicate maneuver, and it shouldn't be attempted late. There's a lot of drag out with the gear and flaps down, *and* there's downward momentum—perhaps more than half the power could handle.

If you normally make rather steep approaches, you'll know what success looks like and can make the decision before making the commitment to land.

When on a low engine-out final—with gear and flaps down and full power on the operative engine—if you see that the airplane is not going to make it to the runway, there are still things that can be done to help.

Fully extended flaps are usually the highest drag items on a twin—draggier than either the extended landing gear or an unfeathered prop. If this is true of the airplane you are flying, consider retracting the flaps back to the approach setting if the approach is too low. The advisability of this would depend on the trim change and stalling speed increase relative to flap retraction. Some airplanes seem to have terrible sinking spells, others don't. Retracting flaps to "stretch" an approach is used only if you are very familiar with the airplane and have practiced it at altitude.

The landing gear offers an efficient way of reducing drag without so many considerations. There's little, if any, trim change and no effect on stalling speed. I'd retract if low on an actual engine-out approach. If I didn't get the gear back down before landing, I'd land on the belly on the runway rather than in the trees with the gear down.

Engine-out Approach Speeds

What about speed during an engine-out approach? The blue line on the airspeed indicator offers a handy reference, and the approach speed should be at this value until the landing is assured. Then it can be allowed to bleed back to normal approach speed.

As during climb, speed must be allowed to dictate pitch attitude. If, for example, the engine-out approach is too low and nothing has worked to make things better, the only thing to do is maintain a safe speed. Any indication below a minimum safe approach speed, or any decay of airspeed when in the area of the minimum safe speed, is a mandate to reduce the angle-of-attack—even if it means touching down short of the runway. To do otherwise is to risk loss of control.

Touchdown on One

Landing the twin with one feathered is a lot different from landing with both engines running, or with one at a zero-thrust setting. There is less drag, and the airplane will float farther. This shouldn't be of concern. The runway selection should have been good enough to allow some leeway, and even with some extra float, a proper approach would have the airplane on the ground by the one-third mark.

Devil's advocate for a moment: Consider a go-around on one engine toward the end of an approach. It's a very risky business, but the problems involved are worth analyzing. Using our 340A again, start with the right engine shut down, 300 feet above the ground, sinking 700 feet per minute, airspeed at 91 knots, headed right toward a good touchdown point, pilot planning the flare and landing. Then a herd of elephants congregates in the middle of the runway. A quick look shows no alternative, no taxiway on either side of the runway to land on, no grass area, only ditches full of starving alligators on each side of the runway. The pilot decides that a go-around is the only option.

First, full power on the operative engine. Maintain 91 knots. What next? Most pilots would instinctively reach for the landing gear, but that's not always best. For one thing, there's usually more drag to the flaps. For another, if the elephants moved and the airplane sagged to the runway while transitioning to the go-around, at least the wheels would be there for rolling.

Flaps should come up to the approach setting. Then if it looks as if the downward momentum is being arrested, gear up. Next, milk the flaps the rest of the way up. The target airspeed then becomes best single-engine angle-of-climb speed, one we haven't really discussed yet. It's 95 knots, which gives a little better climb than 91 knots but also requires some acceleration. That is hard to come by, which explains why 91 instead of 95 is the obstacle clearance speed on takeoff. The distance flown level while accelerating from 91 to 95 doesn't do as much for you as climbing at 91. So, on our go-around, no less than 91 knots, with 95 preferred as soon as some distance can be spared for acceleration.

Cleaning a light twin up and going around from 300 feet might or might not be possible, depending on load, density altitude, and technique. Maybe we made it over the elephants, maybe we didn't. We should have chosen an airport that offered more options.

Single Versus Twin IFR

One of the reasons people buy twins is because they believe that twins offer more options for IFR operation. In a single, an IFR engine failure means an immediate descent through other traffic and the clouds, with an impromptu landing on whatever is available on breaking out of the clouds. Pilots who fly singles and worry over engine failures minimize any perceived hazard by planning IFR flights over areas and in weather that will allow a good shot at a forced landing. The twin seems to offer better odds. It'll stay up after the failure of an engine and should be capable of making it to an airport.

Despite the capability of the twin, a lot of them don't make it to the airport after an engine fails. We think most about engine failure in the critical seconds during take-off, but the engine failure that most often gets the best of twin pilots is the one that occurs en route, and this is especially pronounced in IFR conditions. And consider-ing all the possible times of engine failure in a twin, remember that a perfectly executed approach must fol-low every failure, regardless of when it occurs. Handling the actual engine failure properly only gives you a chance at the approach and landing, which can be even more difficult in serious IFR conditions.

A twin-flying acquaintance used to rib me terribly about flying IFR in singles, which I've been doing since 1955. "Not safe," he said. "Better get a twin."

He used to worry me because his theory was that having a twin meant safety, and while he was instrument rated, he sure didn't practice much.

One day I challenged him to show how much safer his twin would be than my single after an engine failure. We went out in my airplane first, and as I maneuvered for the approach, he pulled the throttle back. The deal was that I could take the hood off when 500 feet above the ground, to simulate a 500-foot ceiling. At that alti-tude, I saw that he had not been kind in selecting a spot. But there was a clearing ahead, and he admitted that I would probably have been able to get the airplane on the ground in the clearing and roll out into the woods. He also agreed that, if I had been able to run the fuselage between trees and use the wings to stop, we'd not have been bloodied.

His turn.

Once off and up, I pulled one engine back to zero thrust. He was doing rather well through the procedure turn on the ILS, but inbound he started lagging behind the airplane. And when he extended the gear at the marker, the airplane promptly sagged below the glide-slope. He was soon 10 knots below the blue line, margin-ally above V_{mc}, and settling below the glideslope. The localizer needle was straying. That was still okay, as he had the same imaginary 500-foot ceiling we used in my single.

He never thought to retract the gear to reduce drag. The airplane was considered out of clouds at 500 feet, when it was quite far off the localizer and below the glideslope. He could see the airport when he took the hood off, but he wouldn't have been able to see it with a mile visibility. It didn't matter, because with full power on the operative engine he started using elevator to con-trol altitude and keep the airplane 500 feet high until closer to the airport. The airspeed went into a steady sink, and direction control problems began. I then gave him the other engine back and we returned to the air-port. Had it been a real situation, he would likely have hit the ground at a high, nonsurvivable rate of sink a mile or two from the airport—after losing control of the air-craft at 500 feet.

He could, of course, have made it to the airport using better technique.

Beggars Should Be Choosers

Picking a good airport for an engine-out arrival is a primary area of technique. This applies not only to runway length, which we've already considered, but to weather and to electronic aids. I would avoid an engine-out approach at an airport close to minimums, for example. I would also avoid an airport whose weather is not reported unless there was every indication that conditions on arrival would be very well above minimums. The best bet would be an ILS airport with a nice long runway and weather comfortably above minimums.

One way in which IFR approaches differ from VFR is that you usually have to go to lower altitudes earlier to find the airport. On an ILS, the glideslope is usually shallower than a normal VFR daytime approach slope, and there's no way to store the energy of altitude. The only thing that you *can* store is speed, and in an actual engine-out IFR approach I'd use this to every advantage. I'd be careful about using the flaps and landing gear, too, and might not use flaps at all if the chips were really down. The drag of the gear can be used to slow or descend, and once the gear is down, it can always be retracted to flatten the descent. Flaps can be used in somewhat the same manner, but the stalling speed and trim changes that accompany flaps might be bad IFR news.

While it's basic to try to avoid a missed approach during an arrival, you don't want to close all doors any sooner than necessary. Storing speed and using drag sparingly can help if the decision for a missed approach should come. Starting the missed approach with only the gear down and 10 knots extra speed would be wiser than starting it with flaps dragging and airspeed at its slowest safe value.

On a big airport with an ILS, I'd fly an engine-out light twin down the glideslope with the flaps up, gear down, and the airspeed about 10 knots above blue line. With runway in sight, I'd start letting the speed bleed back to blue line, and then to normal approach speed as the end of the runway passes beneath. Use of flaps when the landing is assured is optional, but with a long runway it would be best to land in the configuration used for the approach.

On an airport without an ILS, I'd fly the approach with the airplane clean, airspeed 5 to 10 knots above blue line. With runway in sight, I'd intercept a normal VFR approach slope to the runway and extend the gear and the flaps as necessary to track to the runway.

The Pits IFR

The worst possible situation is an engine-out approach to minimums. Here the requirement is perfection. If it's an ILS approach, the airplane has to remain on the glideslope, and at 200 feet the pilot has to look up, see the runway, *and* be in a position to land. Or he has to execute a missed approach from 200 feet. The airplane has downward momentum, the gear is down, and in general there's not a lot working for the pilot on a go-around in a light twin. Given a choice, I'd opt for a VOR or NDB approach to an airport with conditions a few hundred feet above minimums rather than an ILS to minimums. The reasoning is simple: The critical point on the nonprecision approach would occur with more altitude than the critical point on the ILS approach. And an ILS approach is more demanding than a nonprecision approach that will get the job done.

Wherever you happen to be, and whatever conditions might exist, salvation in engine-out situations depends on careful planning and flying proficiency.

Getting to Know Twins

Engine-outs are rare in modern twins, and flying them is normally very enjoyable.

If your first twin outing is in one of the larger and faster ones, the initial impression will be one of great speed on takeoff. In a single, we fly away when at a safe margin above stall, usually at about 60 knots. In a twin, the liftoff speed might be 50 percent faster. The stalling speed is a bit higher in the twin, but the safe single-engine takeoff speed and performance speeds actually dictate when to lift off. In a 340A, we could use 15 degrees of flaps and lift off at 80 knots; instead, we use no flaps and run to 91 to have the best advantage in the event of an engine failure.

Acceleration is good even in smaller twins. The Duchess, for example, will accelerate in standard conditions to 71 knots in 1,000 feet at gross. A Cardinal RG takes almost that distance to accelerate to 58 knots. Simply, the airplane having the lower weight-to-power ratio —it is 10.8 pounds per horsepower for the Duchess and 14.0 pounds per horsepower in the Cardinal RG—will accelerate faster.

The primary reason for not using flaps for twin take-off is that they cut the single-engine rate *and* angle of climb. This is not true with airplanes with excess power, but it sure is on light twins with an engine out. However, if you have to get a twin out of a small airport and are willing to accept a high level of risk in case of engine failure, most will get off in a very short distance if flown with flaps and below safe engine-out speeds. Beech used to advertise that the C55 Baron would clear a 50-foot obstacle in less than 1,000 feet, which indeed it would do. But the airspeed over the obstacle was below both the power-off stalling speed and V_{mc}, and a hiccough of either powerplant would probably buy the farm.

Twins get off so quickly with flaps because the propeller slipstream is blowing over the wings and flaps, while on a single the slipstream just massages the fuselage. Some light twins seem to get light on their feet early in the takeoff run, even though at rest they don't appear to sit at a positive angle of attack. A Duchess is like this. When you reach the recommended 71-knot liftoff speed, the airplane feels eager to fly, as if you ran it for too long on the ground. That 71 is only a knot above the airplane's power-off stalling speed with flaps up, but it is more knots above the power-on stalling speed, which is what counts as long as full power is being developed.

The slipstream effect also can make a difference on landing, especially if you drag an airplane in with a lot of power and then abruptly chop it over the threshold. Doing so in some airplanes can cause quite a sinking spell.

When the Cessna 310 first came out, in the fifties, this was one of the checkout pilot's admonitions: "Don't chop the power when you come over the end of the runway, for she'll fall right out from under you." I'll never forget my first landing in a 310. I had given so much thought to the sink phenomenon that I retarded power very, very slowly after crossing the fence. And I floated halfway down the runway. I've since found it best to make a relatively steep approach in airplanes that have this characteristic. Without much power to chop, there's not much problem.

There can be a little illusion of any airplane falling out from under you when power is retarded. This is caused by the airplane's instinctive nose-down reaction to a power reduction. It wants to seek its trim speed. Most twins have downsprings, thus a rather heavy tug on the wheel is required to keep the nose from pitching down. If you don't pull, the downspring gives you the feeling that the airplane has fallen out from under you. Actually, it has just banged down hard trying to maintain its trim speed—a very natural thing for an airplane to do.

Systems

Since twins often have much more complicated systems than singles, the checkout is more important. There have been instances of pilots not understanding the fuel system on a twin and winding up totally powerless, which is an extreme penalty for misunderstanding the system. For example, if the Cessna 310 is flown on the main tanks until fuel exhaustion and then the fuel selectors are moved to the auxiliary tanks, the chances of the engines' starting in time are not good. Everything in and about the airplane emphasizes the use of the main tanks for takeoff and landing, which means you can't run the mains dry and still land (successfully) on the auxiliaries, but at least one pilot has done it incorrectly. A pilot who is used to a Piper Aztec F, in which any tank can be used for takeoff and landing, might find a problem with a 310 unless the checkout covered systems thoroughly.

Twins: Their Relative Merits

In reviewing what I've written about twins over the years, I gather that a lot of pilots have come to the conclusion that I am antitwin. That is not true. My interest is in being realistic about the assets and liabilities of light multiengine airplanes compared to singles, and indeed they both have assets and liabilities.

On the twin's plus side is its excellent rate of climb on both engines and, in a larger twin, a larger and more comfortable cabin than is possible in a single. The twin's ability to keep flying after an en route engine failure is also a plus, but an examination of the safety record shows that this is often nullified by poor pilot technique. Comparing contemporary singles and twins, there are proportionately more catastrophic twin accidents after en route engine failures than single accidents for the same cause. Some contend that this isn't the fault of the airplane; however, it happens in real situations with real pilots and it must be considered.

An even bigger plus for twins is their redundant electrical and vacuum systems and the ability to carry deicing equipment and weather radar. The singles are chipping away at this as more carry radar and deicing, plus auxiliary means of generating electricity. But it's doubtful that a single will ever match the systems redundancy of a twin.

On the minus side for twins: Often there are two engines out there because the airplane *needs* two engines. In other words, they do not have redundant *power.* When one of its engines is down, a twin will shift from the rocketlike climb of, say, a 9-pound weight-to-horse-power ratio, to the lethargy of an 18-pound weight-to-horsepower ratio. And if the twin is flown correctly—that is, in a manner that always maintains the ability to complete the flight on an airport—then there are often temperature, weight, or airport restrictions on its use. There are many times you just wouldn't fly some twins at gross weight over mountainous terrain if committed to an operation that demands engine-out capability. The fact that there are two engines to fail must also enter in, along with the requirement for excellent performance on the part of the pilot. Perhaps the latter, combined with conservative operating practices, is the key. And pilot performance is determined by proficiency, a product of practice.

7

The Art
of Practice

Nothing, not even practice, makes perfect, but practice
sure can help us fly better.

To be of real value, practice must cover things that
we don't ordinarily do with the airplane. A pilot who
drones from here to there IFR on a regular basis doesn't
need a lot of straight-and-level hood time. A pilot who
spends most of his flying hours doing aerobatics probably doesn't need steep turns. And a pilot who gets his
kicks out of shooting landings need not spend much time
on touch and go's. Each must honestly seek out the skills
that need improvement through practice. That straight-
and-level IFR pilot needs steep turns and unusual atti-

tudes, the aerobat needs hood flying, and the landings specialist probably needs some cross-country work—and the pilot flying a twin needs engine-out experience. Having just left twin operations, let's start with a discussion of practicing in twins.

First, realistic productive training in multiengine airplanes involves some risk. The airlines learned this the hard way years ago, and an accident involving a large four-engine jet illustrates how and why.

The flight was at night, not unusual for airline training flights. A captain and flight engineer trainee, an instructor up for a proficiency check, and an FAA observer were on board. The weather was good, and the crew asked the tower for permission to take off, circle, and land to the north. This was approved.

As the airplane reached V_1 (decision speed), the instructor pulled one of the four engines back to simulate an engine failure. The takeoff continued with no problems. As the aircraft reached an altitude of 1,200 feet, the instructor pulled another engine, on the same side as the first. The trainee was then flying a four-engine jet on two engines—a situation not unlike flying a light twin on one engine.

The aircraft moved around the traffic pattern at an altitude ranging from 900 to 1,200 feet. Headed east, on base leg, the flaps were lowered to 25 degrees. The instructor was prompting the trainee during this time: " . . . don't get below 160 . . . ball in the middle . . . whatever it takes . . . put'er in there now . . ."

As the pattern progressed, full flaps were extended by the instructor without command from the trainee. Shortly thereafter the aircraft descended through 650 feet, airspeed 165 knots. From that point, a 2.5-degree descent slope would have resulted in a normal approach to the runway. There was optimism in the cockpit: "Okay . . . looks good . . . now we're straightened out . . ." Unfortunately the actual descent angle was *three* degrees. The aircraft was headed for a point short of the runway. The trainee apparently sensed this and started a correction, but it was with attitude rather than power. The need for power was later recognized, and was applied to the operating engines. A few seconds later, the aircraft started veering to the left, coincident with a sharp reduction in airspeed and, initially, a steep rate of descent.

The instructor called out the airspeed at 135, and six seconds later said, "See, you're letting her get . . . 'ut the rudder in there . . . you're getting your speed down now, you're not going to be able to get it."

The trainee replied, "Uh uh," and then shouted, "Can't hold it . . ."

The instructor's response: "Naw, don't—let it up, let it up, let 'er up, let 'er up, let it up!"

The aircraft crashed eight seconds later, clearly out of control. The aircraft was in a 14-degree descent angle with a 50-to-60 degree angle of bank at ground contact.

The airlines learned from a series of accidents that maneuvers such as approaches with half power were hazardous in practice, and such things have long since been practiced in simulators. Business jet pilots have the advantage of simulators for this kind of training, too. Unfortunately, those flying light twins must do it in the airplane.

We do have certain advantages over jets. With an engine throttled, the rapid power response of a piston engine could easily save a situation like the one just related. Also, light airplanes accelerate more rapidly, either in response to power or reduction in angle-of-attack, and it is possible to work out of most difficult situations as long as recovery starts *before* control is lost. In a heavy airplane, trouble is probably guaranteed at some earlier point, when the combination of decaying airspeed, pitch attitude, and power setting results in ever-increasing downward momentum.

That doesn't mean you can't wipe yourself out practicing in a light twin. Many people have done it—enough that the mistakes have set a pattern and can be used to develop defenses against the common hazards.

Practice V$_{mc}$ Safely

More people have been lost in light twin training and practice while demonstrating V$_{mc}$ (minimum engine-out control speed) than in any other maneuver. These were the classic accidents during the sixties. The instructor (or FAA inspector in at least one case) would have the student fly the airplane at low altitude and low airspeed, to simulate an engine-out right after takeoff. (The low altitude was used so that airplanes without turbochargers could develop as much power as possible and V$_{mc}$ would be as high as possible.) Then it would be full power, a steep climb, and the instructor or inspector would fail one engine. If the student was really quick and expert at reducing angle-of-attack, and at reducing power on the operative engine if directional control was being lost, the maneuver was deemed successful. The risk came when things *weren't* done correctly. A lot of the airplanes entered low-level spins from which recovery was impossible. The FAA was at fault for prescribing a lethal training maneuver, but I always thought that instructors had the prerogative of culling anything that was apparently unsafe. And I do know that a lot of instructors and FAA inspectors refused to follow the suggested procedure. The FAA finally altered the multiengine flight test guide to eliminate any suggestion of low-level V_{mc} demonstrations. But it was too late for a lot of good people.

If it can't be done that way—in theory, the best way —how else do you practice and prepare for avoiding a loss of control due to asymmetric thrust?

V_{mc} is, among other things, a measure of the airplane's rudder power, which counters the yawing moment created by asymmetric power. If the airplane is not turbocharged and you set out to practice V_{mc} at a safe altitude, say 8,000 feet, there's limited power available for the operating engine and there's little chance that loss of control will occur before the airplane stalls. Even at altitude, however, it is always risky to stall a twin on one engine, because the yawing moment induced at the stall could prompt the airplane to spin.

One solution is to slow the airplane on one engine and maintain control with rudder until the indicated airspeed is at V_{mc}. Then stop applying rudder and continue to slow the airplane. Directional control will effectively be lost and the recovery procedures—a reduction in angle-of-attack and a reduction of power on the operating engine, if necessary—should be apparent. Again, do not stall a twin on one engine.

Keeping One-up on Your Twin

What if you are practicing V_{mc} up high and the airplane *does* wind up in a spin? First, congratulations. You
have just become an experimental test pilot. There is no
FAA requirement for light twins to be spun in the certification process, and most are not spun even by the manufacturers.

An important first thing in a spin is to *cut all power*.
Power tends to flatten spins. Asymmetric power on the
outside of the spin might be an even more notable con-
tributor to flattening than symmetric power. Flat spins
are all but impossible to recover from, so avoiding the
maneuver is the only cure. Most of the airplanes in V_{mc}
accidents hit the ground in flat spins, an indication that
power was probably not reduced, or not reduced quickly
enough. Witnesses to these accidents often report that
power was being developed to impact.

If the airplane is in a spin, and if power is completely
off, prompt application of standard spin recovery techniques might work. Apply full rudder opposite the direction of rotation, and then, about a half turn in the spin
later, apply forward elevator pressure briskly. Neutralize
rudder when the rotation stops, and pull smoothly out of
the dive. Jerky or uneven control application at the moment the airplane stops spinning could result in another
spin, usually in the opposite direction. But don't let it
happen, and it need not; it's possible to fly twins forever
without exposing yourself to the hazards of a stall and
possible spin.

I do most of my twin practicing around the airport.
The engine failure on takeoff is the most talked about

event, and though it seldom happens, the procedure is worth a lot of practice. Regardless of where and when an engine fails, a landing on one engine *must* be made, so this is also worth a lot of practice. There is some risk involved in practice around the airport, but to me the risk is worth the taking, as the place to have your first experience with things like engine-outs is in a carefully controlled practice environment. For example, I like to shoot approaches under the hood with one prop feathered, and land with the prop feathered. I'll do it only with an instructor who is really sharp on the airplane, I'll do it only at a big airport, and I'll do it only in an airplane with unfeathering accumulators. That helps to reduce the risk. There's still some risk, but I prefer it to shooting an instrument approach to minimums at night with one engine out, a bunch of trusting passengers in the airplane, and me never having made an approach and landing with one feathered. "Well, folks, it's like this, I've never done this before, but I am sure it will work out okay."

One of the primary purposes of practicing engine failures during initial climb is to learn to tell the possible from the impossible. If the airplane is flown properly, and has accelerate/stop or accelerate/go capability for the runway, it should be able to continue after passing the decision point, but unforeseen factors can change this. Downdrafts or turbulence can be a factor, or perhaps the engine is not putting out quite as much power as it was when the factory developed the numbers. A lot of things can shoot you down there, and being able to perceive them in time is a life-saving ability.

It's hard to go along with feathering propellers on takeoff in most practice situations. I've done it, but in such carefully controlled conditions that the only real risk was a belly slide on the airport or in an open field. But having done it, now I know exactly what it feels and sounds like—and I know how the airplane behaves *until, as,* and *after* the offending engine's prop is feathered. However, pulling back to idle and then going to zero thrust is a close enough approximation, and it leaves important options open.

I practice landing with one prop feathered, but a lot of people feel that this shouldn't be done. I practice it for the reasons already stated: to make things better if it happens on an actual flight with a lot of other factors in need of attention. That rationalizes the procedure without recommending it.

Speed control and energy management are the keys to engine-out proficiency and the things to work on in practice. Look at the seamy situations, too. With one throttled, not feathered, get the airplane too low, too far away from the airport, gear and flaps down, and examine your ability to maintain a safe airspeed—even though it results in the airplane settling toward a point short of the runway. Then try getting rid of enough drag to make it to the airport. Can you recognize the necessity for a reduction in angle-of-attack any time the airspeed is below a safe value (usually the blue line), or if the airspeed is decaying steadily and is approaching that safe value? If so, you're a safer twin pilot.

In my view, a pilot flying a twin is not current unless he gets a concentrated dose of proficiency flying every six months or so with someone who is truly talented at administering such a program. Without this, the twin is probably a lot more hazardous than a single following engine failure. With it, the opposite should be true.

When and Why Practice Pays Off

When outlining a practice program in any airplane, do so in search of challenge. I usually renew my flight instructor rating by flying with an inspector, because we go out and do things that I don't normally do with an airplane. It's different, and challenging. Turns about a point, for instance, might seem a mindless maneuver until you realize that this drill will help on a circling approach on a rainy night—which you can't really practice.

In turns about a point, the emphasis is on minding the store in the cockpit while keeping up with the ground at the same time. This is also true when groping for the airport in rainy darkness. Give it some thought: It's possible to relate a lot of basic day VFR maneuvers to hairier situations.

Approaches at higher-than-normal approach speeds are worthwhile practice, too. At busy airports, they often ask us to keep the speed up as long as possible. Then, as the standard story goes, the controller asks if we can make the first turnoff after landing. A lot of pilots answer that they can do one or the other—but not both. That doesn't have to be the answer in many airplanes, because we *can* usually do both. It's knowing where to start slowing down to have the speed bleed back to normal approach speed as the airplane crosses the end of the runway. The only way to learn this is practice.

In my Cardinal RG, I'd fly approaches at 125 knots with the gear up. The airplane would slide down the glideslope nicely at that speed, and when on about a mile final, I'd extend the gear (125 being the gear speed). It would then take a lot of additional power to keep the speed up at 125, because the RG's gear had a lot of drag. As the runway got closer, approach flaps would be extended and the power would be reduced to get the airspeed back into the full-flap range. I practiced this until I could keep the speed on 125 and then reduce it smoothly to the recommended 65 knots over the fence. And I used the procedure many times at big and busy airports. Practicing landings with different flap settings will also pay off.

Review and practice (to a point) the emergency procedures in the operating handbook. I don't actually go through *all* the procedures—the insurance company wouldn't like it if I practiced my ditching—but at some time on a flight I'll take a moment to review an emergency procedure in the book. This is a good, often valuable habit to form.

When the Piper Aztec first came out, three friends and I flew one from the East Coast 650 very nautical miles out to Bermuda. We obtained the proper survival gear, including life vests and a raft. The raft was a seven-passenger model, larger than we needed, but the only one we could borrow from the U.S. Navy.

We talked a little about emergency procedures before leaving, but the trip down was in cloud almost all the way, and being out of sight, the water was out of mind.

Things changed on the return. We were flying low to minimize the effects of a headwind. I was flying and another pilot was in the right seat. A couple of hours out of Bermuda, I could tell he was in deep thought and I finally asked him what he was thinking about.

He answered, "Well, Richard, I know that we aren't going to have to ditch, but I was thinking about what I might do if we have to land in the water. First, I'll open the door just before touchdown. Then, after the airplane stops, I'll get out on the wing and open the baggage door with the key. I'll pull the raft out on the wing and inflate it, being careful not to tear anything. Then the rest of you can come out, get in the raft, and we'll shove it off the wing and into the water."

I was slightly amused at the clarity of his vision. The ocean below was churning mightily. The waves were really big, and I had to shatter my companion's idea by saying "Bill, I've never ditched an airplane and neither have you. As big as those waves look, though, the stop might be sudden. And I'd suggest that the first thing you do is duck, because when we hit, the raft will probably go right out through the windshield."

He was quiet on the subject for the rest of the trip. But at least we both had our own idea of what might happen in the worst possible situation, and that in itself would help us avoid surprises.

A more likely thing to think about is a landing gear malfunction at a difficult time.

When most of us are IFR, we use the landing gear of a retractable to establish the final descent at the final approach fix inbound. Prudence suggests that we occasionally pause to review procedures if the green lights *don't* go on signaling that the landing gear is down and locked.

A wide-body jet was once allowed to fly into the ground while the crew pondered and worked over an unsafe gear indication. Another airline jet was landed short of the airport, out of fuel, after the crew had spent too much time fooling with an unsafe gear indication. Still another big jet crashed into a mountain while the crew concerned itself with a similar gear situation. There's a great deal to be said for going through the drills of deciding how you would handle an abnormal gear indication. Two choices: Go on in and land on the unsafe gear or go to a safe altitude and position to work out the problem.

To Spin or Not to Spin

Should we practice spins? It's a popular debating point. Some are adamant that airplanes should have brisk stall characteristics, should be capable of spinning, and that pilots should do spins in training and practice. Others say that airplanes that are easy to spin intentionally are equally easy to spin accidentally, and that airplanes should be spin-resistant—and that there really should be no need for spin training.

I favor the latter position. It's easy to prove that airplanes approved for spins have a higher incidence of stall/spin accidents and that docile airplanes at the stall (such as the Cherokee 140) have a lower incidence. However, in flying an airplane with very brisk stall characteristics—one that is approved for spins and enters them with little effort—I think that spins might at least be experienced, if not practiced.

In an inadvertent spin, it is assumed that the pilot totally failed at controlling the angle-of-attack of the airplane. Rather than spending a lot of time practicing spins, wouldn't it be better to spend the time analyzing the situations that lead to inadvertent stalls and spins? Tight turns at low altitude are the number-one cause.

There's no safe way to practice the low altitude part, but you *can* practice the tight turns. It should be done at altitude to see and feel what happens as the airplane is stalled in a turn. Skid it a little—no perfect ball-in-the-center stalls allowed. If the airplane tends to spin, the measure of success is in how quickly and smoothly the angle-of-attack is reduced and the maneuver is stopped from becoming a fully developed spin. Airplanes that will tuck under from a stall in a turn and start a spin will often respond to a reduction in angle-of-attack and fly right on out without loss of much altitude. Practice this if your machine is placarded as okay for spins—but take an instructor along with you.

The carefully orchestrated stall series is okay as a jump-through-the-hoop training maneuver, but in practice perhaps it is more valuable to experience stalls in as many different ways as possible. Stall it with the ball in the middle as well as with the ball to one side. Don't concern yourself with the stall as such—worry more about recognizing the onset of a stall and recovering promptly. In light airplanes, the instinctive reaction to develop is a reduction in angle-of-attack. If you tend to move the wheel or stick forward a bit at any time control or airspeed seems in doubt, that's fine. But if you try to arrest a rate of descent by pulling back on the wheel, that's *not* a good instinctive reaction. In the jet accident described on pages 156–57, the pilot's reaction to an approach slope that was leading to a point short of the runway was to pull the nose up without adding power. It didn't and doesn't work.

A lot of us tend to go on the defensive when we go out to practice. It could be likened to reluctance to seeing a doctor for fear he'll find something wrong. Any

practice is healthy, though, and practice that finds a
weakness or area of doubt is best of all. Find it, carefully
turn it over in your mind, and strive for precision before
leaving the practice area. Practice on every flight. You
don't have to make a short-field landing at every airport,
but it's free practice to make one at the end of a trip
regardless of how long the runway might be. The same
goes for spot landings. You never *have* to make one in
normal flying, but if the need for a forced landing ever
arises, you'll know what to expect.

8

Emergencies

Emergencies are *unforeseen* occurrences, and strict interpretation suggests that many situations that people consider emergencies really are not. One might even say that the things covered in the emergency procedures section of the pilot's operating handbook are misnamed, because their very presence in the book indicates that someone foresaw the possibility of the occurrence. Any way you slice it, a pilot can perform better in an emergency if a problem is approached in as normal a manner as possible. The mental state of emergency is often confused and stressful, and in studying general aviation accidents I've noted many in which the pilot panicked and, even though all was not lost, effectively gave up.

Here we see the considerable difference between professional and nonprofessional pilots. In reading voice recorder tapes of airline accidents, there's seldom any shouting in the cockpit, even after a crash becomes inevitable. There's silence, a poignant "I love you, Mom," an optimistic "brace yourself," or a four-letter word of resignation. They do keep flying, right to the last second. On the other hand, a lot of general aviation pilots have proved to be screamers in time of distress—showing that they quit flying in advance of the actual accident.

Perhaps one of the reasons that airline pilots fly much more safely than general aviation pilots is that they are more continuously aware that if everything isn't done correctly an airplane can kill you. And when the time comes, they acknowledge it either in quiet dignity or with the profanity reserved for self-chastisement after a grievous error. General aviation pilots often give too little thought to the possible and the impossible in airplanes, and as a result, a manageable situation becomes difficult to handle. While the airline pilot's buildup to a problem is often related to known facts, or has at least been anticipated, the general aviation pilot with a developing problem is usually moving into the unknown. The experienced pilot might feel apprehension, where the less experienced pilot will be affected by growing fear. Fear is a difficult emotion to deal with, especially in an airplane.

A high percentage of the general aviation accidents are instantaneous developments where after making mistakes a pilot winds up in a situation that, even with a

burst of brilliance, cannot be salvaged. There's no way
in which these can be considered emergencies that a
pilot could handle. Some examples are flying into the
ground after going below the minimum descent altitude
or decision height on an approach, or spinning in after
a foolhardy buzz job. In most accidents involving continued VFR flight into adverse weather conditions, the
pilot often doesn't know that he's in deep trouble until
the situation is nearly beyond his ability to respond. A
pilot who hits something on approach at night, or who
flies back into the ground on an IFR takeoff, probably
never knows what the airplane hit. Those are not emergencies.

Hands-on in an Emergency

One kind of real emergency is one in which the pilot
must make decisions quickly and take immediate action
to prevent the airplane from crashing.

An airline accident that occurred when a thunderstorm cell was penetrated on final approach offers an
example of this. It was apparent to the crew that a thunderstorm would be a factor in landing, but the aircraft
ahead had landed without incident. Conditions appeared
suitable for landing, and nothing to the contrary was
reported by a controller as the flight continued toward
the airport.

Three seconds after the flight passed through 400
feet above the ground, the crew heard the tower report
the wind from 210 degrees at 35 knots. That value combined with runway alignment to exceed the allowable

crosswind component for the aircraft, and three seconds later the captain announced the decision to go around. The captain later said, "I was on the verge of going right there, just by looking at the things. And when the tower gave me this wind shift—that's enough for me, I'm leaving."

Whether or not the crew had any feeling of apprehension at this time is open to question. As things happened so quickly, perhaps they did not have time.

Power was applied and the go-around mode on the flight director was selected. The captain called for 15-degree flaps, the copilot complied, and just 16 seconds after the aircraft had passed through 400 feet on the way to the runway, the captain was far enough into the go-around to call for the landing gear to be retracted. As the airplane entered the thunderstorm cell, the airspeed started to drop and the captain lowered the nose to maintain the reference airspeed for a go-around. Eleven seconds after the captain had called for the gear to be retracted, the copilot said: "Pull up, pull up, pull . . ."

According to the captain of another air carrier flight holding on the ground, the aircraft emerged from the rain about 75 to 125 feet above the ground. He said that the aircraft was making a go-around; the landing gear was up, the wings were level, and it had about a 10-degree nose-up attitude. Then, he said, the aircraft appeared to stop flying and descend to the ground with the nose up. It struck and slid, missing the holding aircraft by less than 40 feet before coming to a stop.

In only 27 seconds, the crew had made decisions and had been called on to do precise flying. I'd call this an emergency situation. Any thunderstorm penetration is, at best, an unusual situation. Close to the ground, it is *more* than unusual. The crew followed procedures as closely and quickly as possible, but their actions weren't sufficient to prevent an accident. Some members of the National Transportation Safety Board felt that the accident might have been prevented if the crew had responded a little differently—perhaps by maintaining the nose-up pitch attitude dictated by the flight director command bars even though the airspeed was deteriorating below the proper reference speed for a go-around. That would have had to be a split-second decision on the part of the captain. Reduce angle-of-attack to stop an airspeed decay, or keep the nose up there in the face of decaying airspeed? The airplane might have flown out of the situation if the captain had maintained the flight-director-prescribed go-around pitch attitude, but his instinctive reaction was to follow the other course. The decision had to be made instantly. I think he made the correct decision. Had he not decreased pitch attitude, he would have risked a stall and the consequences would have been even more serious.

For a little mental exercise on split-second thinking, put yourself in the left front seat of the DC-10 that literally lost one engine at Chicago. We know pretty well how that crew responded to the situation. The airplane had passed the decision speed and both the proper procedure and the instinctive reaction was to continue the takeoff after an engine failure. In this case, though, there was some apparent debilitating damage to other airplane systems as the engine came off.

Using hindsight, one possible chance the crew had to minimize the disaster was an immediate abort of the takeoff. The airplane was airborne, and the hit might have been hard. The airplane might have gone off the end of the runway, or out across the field, and it could have broken up or burned. But it might not have dived into the ground with such force. Somebody, or conceivably everybody, might have survived.

Did the crew think about the abort option? Nobody knows. They certainly were not trained to think about it, and had they aborted and made a mess the National Transportation Safety Board might even have come up with a finding of pilot error; as the situation actually developed the fault will be with the airplane.

In the same situation, would you have thought about aborting a takeoff with knowledge only that something had happened out on the left wing?

There was another possible action that might have minimized or helped prevent the DC-10 accident. The leading-edge slats apparently retracted on the left wing because of damage caused by the engine as it tore loose. The leading-edge slats remained deployed on the right wing, causing an imbalance of lift that would make the airplane want to roll to the left. The asymmetric thrust would also make the airplane turn left. With asymmetric thrust and lift combining to make it roll to the left, it is certainly open to question whether or not the airplane could have been controlled under any circumstance, but the one thing that would have enhanced control might have been more airspeed. Lower the nose below the value called for by the flight director and attempt to fly at an airspeed higher than that specified for two-engine operation. One NTSB member noted this as a possibility, and I wondered if he remembered the board's report saying that the DC-9 pilot in the storm just related might have pulled the nose up above the flight director indication to save the day. Pull the nose up one day and push it down the next. In emergencies, pilots do have to think quickly to stay ahead of Monday morning quarterbacks.

In the DC-10 situation the crew probably followed the prescribed procedures to the letter and into the farm. No criticism is offered, but the rest of us should use their problem to stimulate our thinking about emergency actions and procedures.

And into the Mud

It is better to fly in a manner that eliminates split-second judgments. But there will always be times, emergency or near-emergency situations, when split-second decisions must be made. This is why it pays to study the things that happen to others in aviation, as well as to practice.

I found out how fast things can happen one day in a turboprop twin. The day had been nice for flying and we were winding it up on a narrow strip of adequate length. I was in the left seat flying the airplane on a demonstration flight, and the demo pilot in the right seat was a good friend of mine.

The approach wasn't my best, as it followed a rapid descent and low-level traffic pattern—the procedure at that airport. The airspeed seemed high as we came across the end of the runway. It was on about 105 knots, which was 12 knots fast for the operating weight, but it felt even faster. The wind had been given us as northeast and we were landing to the north.

Based on what I saw, I said, "Don't you think we had better go around?" One procedure we hadn't tried in the airplane was maximum use of reverse thrust, and my friend said to go ahead and land and check out the maximum reverse. Split-second decisions.

Touchdown was early, and as soon as the airplane was on, I applied reverse thrust and maximum braking. Something seemed unusual, but I didn't know what it was.

After the airplane had been on the ground for 500 or 750 feet, I heard a tire blow and felt the airplane settle to the left. Nothing seemed badly wrong; I was still working with the reverse, and I thought the airplane would stop in time. But after the tire blew, the deceleration wasn't what it should have been and directional control was difficult. The left tire was out, but the airplane wanted to go to the *right*. My friend in the right seat could clearly see that I was out of ideas and took the controls. He applied full reverse, and we drifted to the right, off the runway, and into the mud.

The root of the problem was twofold. One, the reverse on the left side was inoperative. I did not realize this, and I apparently locked the left brake maintaining directional control with asymmetric reverse thrust. This caused the left tire to blow, and with no braking available on that side and with reverse available only on the other side, we didn't have enough left to slow the airplane and stop it within the runway. It also turned out that the wind was from the southeast instead of the northeast.

Later we talked a lot about the decision processes we used during the blowout episode. I recalled that it entered my mind for a split second after the blowout to go around from that point. That idea was quickly dismissed for three reasons: my knowledge of an aircraft that took off with flat tires and barely made it on a much longer runway, my total ignorance of how much distance this airplane needed to get off the ground with a flat tire, and the commonsense notion that an airplane with a malfunction is better at slowing down on the runway rather than at speeding up.

Until the tire blew, my friend in the right seat had no way of knowing that things were not going well. The airplane was slowing down, and nothing indicated to him that I was using full left brake to keep it straight and slow down at the same time. Given the short time this condition lasted, and my basic unfamiliarity with the airplane, I did not communicate any problem to him. In retrospect, once we decided to land, the first opportunity to keep mud off the airplane came when I saw that it took a lot of brake to both go straight and slow down. Had I gotten off the brakes and applied full power before the tire blew, a go-around would easily have been possible. But that split-second decision slipped by me and the nose of a fine new airplane was muddied. It happens fast. From crossing the fence to stopping couldn't have taken more than 15 or 20 seconds.

Time and Emergencies

For an extended emergency situation, take an airline incident in which the cargo door blew out and affected the controllability of a DC-10. The airplane was at 11,-750 feet and climbing when the crew heard and felt a thud. Dust and dirt flew into their faces, the rudder pedal moved to the full left position, the thrust levers moved back to near the flight-idle position, and the airplane yawed to the right. The captain said he lost his vision momentarily; he thought they had been in a midair and that the windshield was gone.

Now, here you are, more than two miles above the earth in a big airplane with everything fouled up. What do you say?

One crewmember uttered what the NTSB calls a nonpertinent word, and then the captain asked, "What the——was it?" Someone whistled. The fire warning horn blew, and seven and a half seconds later the crew was actively engaged in trying to figure out what was wrong with the airplane. About 45 seconds after the problem started developing, the air traffic controller was advised of an emergency, and about 40 seconds after that the controller inquired about the nature of the emergency. The pilot replied, "We have a control problem, we have no rudder, got full jam, we've had something happen, I don't know what it is."

The controller cleared them toward the airport, and the crew acknowledged a descent clearance with word that they would be letting down slowly. In less than three minutes they transmitted: "Okay now, we've got it, ah, problem, I got a hole in the cabin. I think we've lost number two engine, we've got a jammed full left rudder and we need to, ah, get down and make an approach." And a moment later: "I have no rudder control whatsoever, so our turns are gonna have to be very slow and cautious."

There was no need for split-second decisions. The airplane was flying and it was controllable. However, the crew knew full well that a difficult situation lay ahead.

Four minutes into the problem, they determined that they could control the airplane in the yaw axis by using differential power on the wing-mounted engines. Some flaps were extended, and as the aircraft descended, one of the pilots said, "We've got a nice rate of descent; even if we have to touch down this way, we're doing well."

The landing gear was extended at the normal point in the approach, as were landing flaps. After the airplane was cleared to land, the captain remarked that he had no rudder to use in straightening the aircraft out after touchdown. They were unable to keep the aircraft on the runway, but the stop was successful and the emergency evacuation was conducted promptly.

When the cargo door blew out, a portion of the rear floor sagged as the cabin depressurized. One cable to the left rudder and a pair of elevator cables were severed. The manual stabilizer trim handles were rendered inoperative, and the rudder pedal torque tube and the left rudder cable horn were separated. The NTSB commended the crew for handling the aircraft well despite these conditions.

In light airplanes we seldom see failures that compromise the controls, but it is possible, and every eventuality should be anticipated. I was flying with another pilot one day, and the airplane was misbehaving to some extent. It was a frigid winter day, and the trim tab of the airplane wouldn't move. The airplane was a little out of trim—in the nose-down direction—but we were able to hold it.

As we droned along, I studied the situation. With operative trim, there's redundance in pitch control. Should an elevator cable fail, the airplane could be controlled and landed, using the trim. With no trim, and with the nose wanting to go down, any failure in the elevator system would be serious indeed.

We talked about it. The first thing to do would be for the other pilot to clamber over in the back, perhaps all the way to the baggage compartment, to move the airplane's nose up. Then a power reduction, or flaps, or the landing gear could be used to make the airplane nose-down. There would be some control. I could make the nose go up or make it go down. The airplane could then be flown to a big airport, where a landing could be attempted using power, gear, and flaps against the basic nose-up attitude.

Having thought it through, I felt better as we flew along. I felt even better after getting the trim fixed at the next stop.

Weather Emergencies

Weather is a frequent cause of "emergencies" in general aviation, and the pilot's mental state has a lot to do with their outcome.

Consider a pilot without an instrument rating who gets into cloud and sees no quick way out. The first reaction might be panic. Even a pilot with an instrument rating doesn't feel at ease after an unexpected penetration of instrument conditions, and the instinctive reaction is often to do something very quickly. This is not the time for split-second decisions. This is the other kind of emergency, like the airliner with the blown cargo door. The need is for a series of well-planned actions that must be accomplished with care.

If something is done quickly in a serious situation that doesn't require instantaneous action, the pilot may barter away one thing in his favor—a little time. I've heard good pilots talk to themselves in time of stress: The right words can be a steadying influence. If suddenly on instruments for the first time, a noninstrument pilot should first describe the situation to himself—he should have a conference with himself on the alternatives.

By taking the time to consider all the alternatives, the old brain will at least be exposed to the best one. If the training has been good, and if some study has been put into analyzing similar situations, there's a good chance the best alternative will be the course of action finally chosen. It might make the difference between loss of control in a hasty turn following a snap judgment and a successful conclusion of the flight.

Calm helps, too. In this example, the airplane is in cloud, so this means that the primary chore is to relax and work at instrument flying. Once mastery of that is established, it is okay to go to the chart or radio, while still not neglecting the instrument flying. It's a matter of deciding what action is primary to keep the airplane flying and of calmly giving that priority over things that might be considered as busywork.

The danger of cockpit busywork at the wrong time was illustrated by a problem the Navy had some years ago with pilots flying into the water after night carrier launches. The same type aircraft was involved in each accident and they knew there had to be a common thread. Finally it was found: An item on the checklist called for the pilot to turn his head for a visual check of something behind him immediately after takeoff. That took the pilot's eyes off the instruments and the head movement helped induce spatial disorientation. When the pilot got back to the instrument panel, he was at a distinct disadvantage. One rule-of-thumb is to minimize head movement in a situation where spatial disorientation is a possibility. Look away from the instruments only in a situation where a momentary deviation from the proper attitude isn't critical.

I saw the head movement phenomenon used another way when riding along in the rear seat while a colleague was taking a check ride for an instrument rating. When it came time for unusual attitudes, the inspector had the applicant close his eyes and bow his head as if to look at his lap. Then he told him to move his head, still bowed, from side to side. Then he put the airplane in an unusual attitude. The applicant recovered but later said he was terribly disoriented during the recovery. The head movement had of course affected his inner ear.

The 180 and Other Saves

If it is possible to get out of weather, the well-advertised 180-degree turn is wisest once the pilot is settled down and resigned to the chore at hand. But not a 180 that is too hurried and too tight due to stress. The pilot who goes on instruments at low altitude in grungy conditions and makes a fast decision to turn and get out fast might lose control while trying to hasten the turn or might fly into the ground. The situation might be one to handle in another manner.

For example, consider the pilot scud-running in very low conditions. It's raining, he's at about 500 feet, and the visibility has gone from poor to worse. Suddenly the airplane is in cloud that the pilot didn't see coming. This is it. On the gauges at 500 feet. What to do?

A 180-degree turn might bring the pilot back around to the poor but visual conditions in which he had been flying—but it might not. The success of the turn would be measured in maintaining control and altitude, staying clear of higher terrain, and in completing the turn and flying into an area where the weather is as good as it was before entering the clouds. *A lot* would have to go right for the conclusion to be successful.

The pilot made a big but common mistake in flying the airplane into conditions where an inadvertent blunder into clouds was possible, and the 180 might be dangerous at this late date. It should have been made long ago. What next?

Pilots have saved themselves in such a situation by gradually climbing to a safe altitude and getting helpful information from an FAA flight service station or air traffic facility. The people on the ground can't fly the airplane for you and can't make decisions for you, but if radar contact can be established, or if the airplane's position can be determined using direction-finding equipment, the controllers can do the navigational chores. This leaves the pilot to concentrate on *flying* the airplane.

One of the worst things a pilot can do in this situation is try to shift complete responsibility over to the people on the ground. In transcripts of conversations between lost pilots and ground facilities, you can at times sense the pilot trying to move his responsibility into the microphone and down to the ground. Pilots ask controllers for things they *know* the controller can't provide. Even after a loss of radar contact, pilots keep quizzing the controller about the best direction to fly, and after moving into turbulence, they beg for advice on smooth air—as if that were depicted in the controller's scope.

A pilot who knows the score uses ATC to every possi-

ble extent but does not try to transfer responsibility.
When a pilot gets in a bind, the best action is to tell the
controller what is needed. "Clear the runway." Or "Give
me vectors back to the airport." A noninstrument pilot
in cloud should confess his predicament and ask for the
best deal. For example: "I am not instrument rated, I am
in cloud, I think I am about 25 miles north of Lynchburg,
I am at 3,500 feet. I would like to go where the weather
is better so I can get below the clouds."

Given that scenario, ATC has something to work
with. An attempt can be made to positively identify the
aircraft, and then it can be handled in much the same
manner as an IFR aircraft being radar vectored. It is far
better than a feeble "Can you help me?"

IFR Weather Emergencies

Move now to IFR pilots, where the most frequent
weather emergencies are thunderstorms and ice. Thunderstorm penetration is an emergency that can and at all
costs should be avoided. But if a pilot flies enough IFR,
chances are he will get into a thunderstorm, or close
enough to one to have a real problem.

The air is incredibly turbulent in or near a thunderstorm, and a pilot's thought processes can suffer to the
point that giving up might be considered as one of the
alternatives. I've heard more than one pilot admit that
the situation seemed completely out of hand and that
there was nothing that could be done. But then came the
realization that the engine was running, the instruments

were working, and the airplane was in one piece. With that much going for you, why quit?

A retired airline pilot, while describing what *might* have happened to a friend of his who flew into the ground in a storm, gave me a realistic picture of the feeling: "You know, the turbulence seems to come in jabs and the airplane vibrates as well as pitching and bucking and rolling. The instruments take this up and it's hard to tell what they are doing. The airspeed is jumping wildly, and no matter how tight the belt is, you are moving in the seat and it is difficult to see the instruments. You start trying just to average things, to keep it all in the safest possible place." He added that his friend just must have flown into the ground while trying to average it all out.

One thing that I would add to his description is that in turbulence and on instruments the first and strongest feeling is one that questions *why* you are in this position. It is, after all, one of the more unique situations in life. When in a storm on a ship, sail can be shortened, sea anchors can be rigged, and there's always the raft if all else fails. In an airplane, it's pilot versus nature with no backups. Anyone who has ever been near or in a storm knows how small, frail, and out of place you can feel.

But storms don't last long, and certainly the pilot who uses the basics of instrument flying in turbulence increases his chance of coming through in one piece. It is a problem that can be handled if everything is done correctly.

Hands-on in Ice

Ice is an emergency that develops more slowly than a thunderstorm, and it can usually be avoided by deliberate thought and planning. The main thing is to get out of icing conditions. There is a strong probability that the airplane was flying without any ice before it started collecting it, and ice-free air is still back there, *behind.* Pilots who fall victim to an ice emergency are the ones who continue so far into icing that they are an unreachable distance from the ice-free area—or until there is so much ice on the airplane that it won't fly in *any* kind of air. Ice should be treated like a fire. If you have one extinguisher, better use it at the first sign rather than waiting to see if the wisp of smoke is going to turn into a conflagration. Likewise, it is best to change plans at the first sign of ice rather than waiting until the airplane feels as if it is about to fall out of the sky.

One Down, None to Go

A frequent emergency in general aviation singles relates to the engine. More often than not, the pilot causes the engine problem by doing something stupid, like using all the fuel. But once silence sets in, the problem must be handled regardless of cause. We've already discussed the handling of twins with an engine out—what about singles?

When driving along a highway or just looking out the window at home, it is often difficult to imagine how an airplane in trouble might be landed successfully. I thought about this while driving along a road in south- east Alabama, which is generally flat terrain. Things looked good for a forced landing for a long stretch; then we passed through a series of deep gullies. What if the altitude and airspeed started running out with just those gullies below and ahead?

As bad as it looks, landing in random places often isn't as threatening to life and limb as it might seem. Three accidents that occurred near my home within a reasonably short period of time illustrate this. I lived in Little Rock, Arkansas, at the time, and while the details of the accidents are hearsay, the causes and the outcomes are known. All three mishaps involved single-engine air- planes, and all were at night.

In the first, the pilot was working his way toward the airport in pretty good weather when the engine quit. Powerless, the airplane started down. He was apparently trying the "black hole" theory of night forced landings. The area was built up, and where there were lights, he probably reasoned, there would also be buildings and obstructions.

He flew the airplane, a Skyhawk, into the black hole at minimum speed without stalling or losing control, and on arrival found that it wasn't too good a choice. In fact, it was an electric company substation. The airplane tan- gled with the wires and lights went off over a large area. The airplane was mangled, but the two occupants were not seriously hurt. In fact, had they been wearing shoul- der harnesses, they might have escaped injury alto- gether. The damage was done when faces hit the instru- ment panel. The cause of the engine failure was later said to be carburetor ice.

The next event came toward the end of a 675-nauti-cal-mile hop in a Cherokee. IFR at night, with three passengers, the pilot had assumed that the airplane would go the distance. He was wrong.

As the airplane was maneuvering for an ILS ap-proach, the engine took the last gulp of avgas and quit. By choice or by chance, when clear of clouds the pilot headed the airplane toward a dark area. This time it was a cypress bog in an area of bottomland and creeks. The Cherokee must have been under control until it mixed with the treetops. The airplane slowed down and finally wound up at the base of one of the trees. Tall trees, and it was a long way down. The airplane was demolished, but none of the occupants was injured. Shoulder har-nesses had been installed, and while I never got definite word that they were in use, the outcome strongly sug-gests that they were.

The third example also involved a Cherokee. The weather was bad, foggy bad, and the pilot passed his original destination and headed toward Little Rock. As he got closer, the tower informed him that the Little Rock weather had gone down in fog. The pilot then apparently headed toward Fort Smith after reportedly telling the tower that he didn't have much fuel on board.

The engine quit shortly afterward—the last of the gas was gone. Foggy night and all, the pilot maintained con-trol of the aircraft until it hit something. In this case it passed between two trees, which removed the wings. The fuselage slithered to a stop, and again the occupant was okay.

There's no way of knowing just how cool and col-lected these three pilots were as they worked toward a night forced landing, but the fact that they retained con-

trol of their aircraft is testimony that they didn't forget
the laws of aerodynamics. That is perhaps the most important thing to remember in a power-off emergency.
Whether the cause is mechanical (rare) or some lapse on
the part of the pilot (common), the outcome depends on
about nine parts basic airmanship and one part good
fortune. The ratio might vary a little, but even in the
most extreme circumstances the airmanship counts for
most.

Another forced-landing example comes from the
mountainous west. The airplane was moving along IFR
when all power was lost. As the airplane broke out of
clouds, a passenger saw that they had just cleared a ridge
at right angles to their path. So much for good fortune.
Finding himself in a deep and narrow valley, the pilot
headed for the creek at the base of the valley. It wasn't
the most appetizing landing site, but it would have to do.
He got the airplane onto the ground at minimum speed;
shortly after it touched, one wing hit a big rock and the
airplane turned and slid sideways until it stopped. One
person was injured, not badly; the other two were okay.

Airplanes have been left in all sorts of places with
varying degrees of success. The key is in flying at a minimum safe speed until the airplane touches and natural
forces stop it.

In studying events like this, it is very clear that more
than passing attention should be paid to survival gear.
We seldom think it will happen to us, but after a landing
in a remote and rough area, good survival equipment
could transform a harrowing ordeal of survival into
something like a camp-out.

It's in the Book

The collection of emergency procedures in most pilots' operating handbooks should be read with some regularity. If the time ever comes, there will be little or no time to review the procedures. And it's also possible to use this information to minimize the severity of a problem or to avoid exposure to unnecessary hazards.

I usually fly across Lake Michigan at least twice a year —going to and from the EAA convention at Oshkosh, Wisconsin. It's about 41 nautical miles across at the lake's narrowest point. Even with flotation gear on board the consequences of an engine problem are much less if the airplane is always at a point and altitude from which it could glide to land. Going across at 12,500 feet makes this possible in many light airplanes; some may need to fly a little higher.

Gliding distance is just one item in the book on my airplane. Fires, ditching, a rough engine, an opened door, and a lot of other things are covered. It's all good reading, worthy of an hour or more a month.

9
Flying at Busy Airports

Many experienced pilots don't like to use big and busy airports. Terminal control areas and radar service are strange to them, and more often than not, things do not go smoothly when they use an airport with a lot of traffic. In most cases this is simply a matter of state of mind.

In my view, Washington National is our "busy" airport of record. It has three runways. The longest runway, 18/36, is 6,869 feet long and is the primary runway for air carrier jets. Runway 15/33 is 5,212 feet long and is used by air carrier jets only when there is a strong north-westerly wind; even then, I've heard some air carrier pilots say that their aircraft is too heavy for the runway. Runway 3/21 is 4,905 feet long. All the runways cross each other. On 21 there is adequate length for a light airplane to land and hold short of the long runway. It's also possible to land a light airplane on 36 and hold short of 15/33, or on 18 and hold short of 3/21.

The runway configuration isn't ideal for high capacity operations, and neither is the airspace around the airport. A prohibited area over the White House and Capitol blocks straight-in approaches to runways 18 and 21 and requires a turn soon after takeoff on 36 and 3. Andrews Air Force Base is about 10 miles east southeast, and airspace has to be shared with that facility. There are also other prohibited and restricted areas in the general vicinity. The rivers nearby help make it work. The Potomac River flows from the northwest right by the airport and then heads south. The Anacostia River flows in from the northeast. These create natural flyways, and two IFR approaches are based on the Potomac. Coming in from the northwest, with a ceiling of 3,500 feet or better and four or more miles visibility, the drill is to follow the river in and make a last-minute, low-altitude turn to land on runway 18. The approach to the north is nicer, right up the river and straight to runway 36. The minimum for that approach is 3,000 and four.

The controllers at National do a super job with what they have. The highest number I've seen for hourly aircraft operations there is 128, on a VFR day, but I wouldn't be surprised if they have moved more. The airline pilots all know the procedures; the controllers seem to sense when a general aviation pilot is familiar and capable and when one might need a little extra help. They handle the traffic well and usually keep the fast and the slow in entirely different airspace until it is necessary to mix them up in the landing pattern.

One thing I've learned about National and other busy airports is to arrive IFR even in good weather. This makes life simpler. Show up IFR and the approach controller takes you on a handoff from a center controller. He knows your number and he sees you on radar before you call. You are in the scheme of things. Show up VFR, call when 30 miles out, and introductions have to be made. Then the controller has to look at his traffic situation and decide where and when you'll fit in. National is one of the few airports that operates with a reservations system—IFR reservations are limited and VFR arrivals are handled as possible, controller's discretion—but if the weather is decent, they always try to accommodate all the traffic that shows up. Again, the IFR arrivals are expected and National does a good job of handling them in fair weather or foul. VFR you must strike a deal on the spot.

I flew an interesting trip into National one late winter day. The weather was clear, but I was IFR. The clearance was over Nottingham VOR, east of Washington, direct to the Washington VOR, which is right on the airport. The assigned altitude was 6,000 until—when I was about 10 miles from the airport—the controller said to descend to 4,000. The airport was in sight at this time. Landings were to the north, with a rather strong northwesterly wind. Runways 33 and 36 were being assigned to landing aircraft, with a few air carriers refusing 33 because of their landing weight.

It's customary at National to deal with two approach controllers on the way in. One handles the initial maneuvering, the second fits you into the final flow of traffic and turns you over to the tower at the last minute. The second controller this day explained our sequence:

"You'll be following an Eastern 727 on a 12-mile final."
He then gave me a heading to fly. I could see the jet and
could sense what the controller was doing. I was a few
miles away from final, perpendicular to the final approach course and pointed at a spot ahead of Eastern's
present position. The jet's speed would move it ahead,
and I'd get there and turn about a six-mile final right
after it passed.

It's tough for a controller to mesh slow and fast airplanes, and I could see one problem developing. It
would all work fine based on present speeds, but after I
turned final, I'd be flying into a strong headwind and my
groundspeed would slow by perhaps 30 knots. The jet
would fly away and leave a big gap. The inevitable following jet would gain on me. I told the controller that I
had the jet in sight and he said to follow it as closely as
possible and call the tower. I said I'd do that, and the
airline pilot chimed in and said, "Close but not too close,
okay?" Okay. I turned parallel to the final approach
course and started to plan it so I'd land a minimum safe
distance behind the jet. I had to start out close because
I could only change things to increase the rate at which
he was moving ahead. No way could I gain on him.

The Cherokee I was flying would do about 140 knots
indicated, and this was working fairly well against the jet.
He was gaining enough to space me for the landing. We
were both given runway 33.

There was another jet in the picture as we neared the
airport. A 727 was behind us on final to runway 36. The
plan was for the first jet to land on 33 and for me to
follow him and get down and through the intersection of
33 and 36 before the other jet touched on 36. If it looked
okay to the controller, it looked close from the airplane.

I remarked to my passenger that we'd probably have to go around. When I was about a quarter of a mile out on final to 33, the jet was over the approach lights to 36. The tower told me to go around and to enter a left downwind for 36.

When I was on left downwind, the controller pointed out a Navajo on final and asked that I follow it, "as closely as possible, please." There was another of the never-ending stream of jets on final. The controller and the jet discussed using 33, but its captain answered, curtly, that he had to have 36 because of weight—and that he'd been slowing and S-turning for the last 10 miles to fit behind the Navajo. It was obvious that he didn't think much of the tower for landing two light airplanes before his turn for the turf.

The tower had it all figured out and cleared me to land.

I tried extra hard and landed right behind the Navajo, rolling into the high-speed turnoff not 10 or 15 seconds after the twin. The jet touched down probably 15 or 20 seconds after I landed. Airplanes come in a continuous string and the controllers handle them well.

If it seems a bit chaotic to read about, it's really quite logical when you are flying. It's mainly a matter of following instructions from the controller, who is referee for both the airspace and the runway.

The key in flying at Washington National, and at any other busy airport, is in knowing what to expect. It's essential to know the airport layout—if a pilot lines up for the wrong runway, or becomes confused over which is 33 or 36 or 3, it breaks the flow and can send a lot of airplanes scurrying for the holding patterns. And if the controller asks you to land and hold short of a crossing runway, and you make the approach too fast in trying to comply with a request to make "best time to the airport, please" and then can't hold short, it could foul up the whole works.

One way to learn an airport is by studying the approach plates, the airport diagram, and the standard terminal arrival routes, if any are published for the airport. Also, have a look at the frequencies. If ground control is other than the standard 121.9, you will remember it. If your radio has frequency storage capability, load the tower and ground frequencies well in advance. Study taxi routes, too; once on the ground, it is difficult to taxi and consult the chart at the same time. When inbound, the automatic terminal information service will have the approach and active runway. This completes the picture of what to expect.

Communications

Radio technique is especially important in a busy terminal area. The key is to be as brief as possible without sacrificing the clarity of your message.

On first contact with approach control, at a TCA airport, a good call might be: "Cessna 40 Romeo Charlie is over New Brunswick, VFR, 2,500, inbound to Newark with information Yankee." Some would add the name of the facility being called, "Newark Approach" in this case, to the first of the transmission, and that is fine, even desirable in many instances. But if you will listen to airline crews, the best communicators of all, you'll hear that they often do not address the controller by facility. They

must feel that having the radio tuned to the proper frequency eliminates any need for telling the controller
what his name is when calling.

Listen carefully when the controller calls you, and if
you do not understand the message, don't hesitate to ask
for a repeat. If you have consistent trouble understanding, review your listening technique. The biggest problem understanding controllers comes when a pilot tries
to understand the message as it is being given and a
question comes to mind during the communication. The
mental question effectively cuts off all said after it and
will usually necessitate a repeat. The best way to listen
is with an open mind, taking in all the words to put
together as a complete message at the end of the transmission.

Watch for confusion. I saw examples of this at controlled airports within a period of a couple of weeks, and
they illustrate how critical it is to be precise and to listen
carefully.

A colleague was flying a Piper Six 300 inbound to the
airport from the north. The airplane number was
N3051U. When he called the tower, he referred to his N
number as "51 Uncle" and said he was eight miles west
—we were actually *north* of the airport. A moment later
another airplane called in from west of the field. The
tower told the pilot of the other aircraft to watch for us.
My pilot then called, using "51 *Uniform*" this time, and
reported that he was north of the field instead of west.
The tower operator took this as an airplane in addition
to 51 Uncle, instead of a correction of position. In his
mind, he had three airplanes inbound instead of two. As
we got close to the field, my pilot called entering downwind. The tower was totally confused by this time, and

for some reason he thought 51 whatever was an airplane just making a missed approach off a simulated ILS approach. When we got that straightened out, there was no further confusion. But I suppose the tower controller is still wondering what happened to that other airplane inbound from the west.

In the second episode, I was joining the pattern at a very busy airport and was told to enter a mile southwest of the airport for a right downwind leg to runway 23. My Bonanza's number was 17838. I heard another airplane inbound with the call letters 83 Yankee, and when I got a word in edgewise and reported on downwind leg, the tower said: "Roger 83 Yankee, continue, you are number two following a Cessna on right base." In reality, 83 Yankee was behind me and what the controller meant was that 838 was number two. I called his attention to the fact that there was an 838 and an 83 Yankee in the pattern and there seemed to be some confusion. He got it straight, and we had no further problem.

There will always be human errors in the traffic control system, and pilots should be alert to them. Stay alert, and say the word if you perceive confusion. I'd say that you'll more likely see these errors at smaller facilities.

The ATC/Pilot Relationship

A visit to a terminal radar approach control facility can be very helpful in understanding the controller's problems.

Traffic appears flat on the radar scope. It is not three-dimensional, and while controllers have altitude information on airplanes with altitude-reporting transponders, they still tend to separate airplanes horizontally on their screen. If two are crossing with only vertical separation, you'll see extra attention to verifying altitude. A lot of horizontal separation might look wasteful until you realize that the inbound airplanes all end up at the same altitude—the field elevation. And at some point early in the terminal area proceedings the airplanes have to be separated horizontally lest two get to the runway at the same time.

As pilots in traffic, we often think we have a better grasp of the traffic picture than do the controllers, and it's common to hear pilots tell controllers how to run the airspace. The controllers at the radar screen have the big picture, though, and there is a valid reason for their every heading and altitude assignment. Watching them work for a while helps to understand both sides of the situation.

Wake Turbulence

The avoidance of wake turbulence is related to operations at both busy and quiet airports, more so to the former because there are more big airplanes there.

The hazards of the wake come from a vortex that forms at the tip of any lift-producing device, with the wingtip vortexes the strongest. Those that form at the ends of the flap trailing edges and at the tips of the horizontal stabilizer also qualify as vortexes, but they lack the strength and staying power of those at the wingtips.

The intensity and size of a vortex is influenced by the wingspan, weight, and speed of the generating aircraft. A heavy airplane flying at a slow speed would generate the strongest vortexes.

A vortex trails back from the wingtip and spins, developing a core with a diameter roughly equal to 15 percent of the total wingspan of the airplane, or 30 percent of the length of the individual wing that is producing the vortex. The rate of rotation of the vortex is related to the speed of the aircraft at the time of generation. Lower airspeeds generally produce stronger vortexes.

The vortexes have a life of from two to five minutes in absolutely smooth conditions, though it has been suggested that they can last even longer in very stable air. Turbulence, from wind or surface heating, tends to tear the vortexes up.

Vortexes generally settle at a rate of approximately 500 feet per minute until they are 900 to 1,000 feet below the altitude of generation. Atmospheric lifting can, however, impede or stop the settling of vortexes. Generally, atmospheric lifting is related to turbulence, so both lifting and preservation of the vortex isn't too common.

When generated close to the ground, vortexes settle parallel to each other until they are from 25 to 50 feet above the ground. Then they start to separate, moving away from each other. The rate of motion across the ground is from 300 to 500 feet per minute. That's about five knots, so a five-knot crosswind component might keep one vortex over the runway after the takeoff or landing. The other vortex would move away from the runway at 10 knots.

The vortex of any airplane that is larger or heavier than the one you are flying is a possible hazard. The vortex of a 727 would encompass a high percentage of the span of a light airplane, and it is logical that it could and would completely override the light airplane's control power. The vortexes from a business jet, while not as large, spin furiously and if encountered in a light airplane might be enough for a loss of control and upset. Even a lumbering old piston airline type can create strong enough vortexes for an upset.

On takeoff the vortexes don't start rolling off the tips in earnest until the nosewheel is lifted from the runway. When the wing is running level, it is not creating much lift, and thus the vortexes are not strong.

The key on takeoff is to lift off before reaching the point where a preceding heavy lifted off, and then climb above the vortex of that airplane.

Lifting off at a point *before* a heavier airplane has rotated is usually not a difficult thing to do. A jet will run down the runway for at least 3,000 or 4,000 feet before liftoff; a lighter airplane is off in 1,000 or 2,000 feet, so if you start from the same point you should be okay.

Climbing at a better angle than a jet isn't so easy. They climb at much higher airspeeds than light airplanes, but the rates of climb can be proportionately higher. If you doubt that you can climb above the path of a heavier airplane, make a turn soon after takeoff. Consider the wind when turning. If the low-level wind aloft is from the right, for example, the turn would best be made to the right. Turning left could take you back into vortexes that had blown away from the extended centerline of the runway.

Beware an intersection takeoff behind a heavier airplane that started its roll from the end of the runway. You might not get off before its liftoff point and you might fly into its vortexes. At controlled airports, controllers won't issue an intersection takeoff clearance for three minutes after a heavy airplane has passed that intersection. You might as well spend that time taxiing to the end. At uncontrolled airports, you are on your own in avoiding intersection takeoffs behind heavy airplanes.

A landing airplane can cause a problem for a departing light airplane. For example, if a 727 lands 1,500 feet down the runway, and the pilot puts the nosewheel down 2,000 feet down the runway, strong vortexes might well be present over the first 2,000 feet of runway. If there's a little crosswind, a vortex could remain right over the runway. When departing in such a situation, either wait a couple of minutes or delay liftoff until past the point where the 727's nosewheel touched the ground. If there is a slight tailwind component on the runway, beware. The vortexes created as the heavy airplane landed will move up the runway.

Wake Turbulence on Approach

On approach, either stay above or avoid the trail of a heavier airplane. Theoretically, if you fly the glideslope when following a heavier airplane on an ILS, you'll be above his wake because of its settling tendency. But what if the pilot of the heavier airplane wasn't precise and flew the glideslope a bit high? If you were following a minute behind, the airplane would have to have been grossly high for its vortex to be able to settle 400 or 500 feet and still be on the glideslope, but the possibility always exists. If following a heavy, I usually fly a couple of dots high on the glideslope to give an extra measure of protection.

Also, a tailwind component on the approach would result in the vortexes remaining on the glideslope, or settling relative to the glideslope at a lower than normal rate.

The important thing to avoid on an approach is getting below the path of the heavy airplane. If below it, you're likely to be exposed to its vortexes when flying to the same runway.

When landing on the same runway as a heavier airplane, plan a steeper approach and land past the point where its nosewheel was lowered to the runway. If landing behind a departing airplane, touch well before the point where it was rotated to a flying attitude.

If following a heavy airplane that goes around, I'd always abort my approach and go around too. The airplane ahead leaves its trail all the way down the approach course and runway when it goes around, and there's no way to get to the runway without flying through it.

Some large airports have parallel runways, and the vortexes of heavy airplanes using a runway parallel to yours should always be considered, especially if there is a bit of crosswind blowing across the runway used by heavy airplanes toward the parallel that you might be using. In this case, it's best to plan as if the heavy used your runway and take off or land accordingly.

Terminal Area Wake Turbulence

Wake turbulence encounters in terminal areas are very possible, and pilots who fly the major terminals have gotten some pretty good jolts.

I've flown through the wake of heavier airplanes three times, each event was several thousand feet above the ground, and each penetration of the wake was at a sharp angle.

When flying into a vortex that's moving in roughly the same direction, the effect is an overpowering roll action. When flying through one at a right angle, the product is a memorable bump. One airplane I flew through a vortex had a G-meter in the panel, and the reading was near the structural limits of the airplane. This prompted a thorough inspection, but no structural damage was found.

In my most recent encounter, I was descending VFR and had been given a 727 as traffic in a holding pattern ahead. I saw the 727 flying toward me from quite a distance, and right after I spotted it, the controller issued a clearance for that flight to descend and turn back toward the destination airport. The pilot turned smartly and started a descent, flying a couple of miles in front of me at a right angle to my path. Based on how it looked, I thought that I'd fly over the vortex, but just in case, I slowed my airplane to its turbulent-air penetration speed.

The precaution was well taken, because I indeed flew through one of the vortexes. Right before I hit it, something told me that it was coming. A vortex can't spin without affecting the air over an area much larger than

the core, and I suppose that before the core was penetrated there was some split-second clue as the airplane
entered the affected area.

In another encounter I was being vectored for an ILS
approach at a major airport when a heavy jet passed
1,000 feet overhead, going in the same direction. I was
to follow this traffic for the ILS, and it was obvious that
it would descend through my altitude—and that I would
be exposed to its vortexes at some point.

The jet was given a turn on base at about the same
time it was given as descent clearance. I had told the
controller I was concerned about turbulence avoidance,
and he had me turn about 45 degrees and maintain altitude. This was right after giving the jet its clearance to
turn and descend. My turn was in the same direction as
the jet, however, and a minute or so later I flew through
the vortex at an angle to it. It was reasonably weak but
still strong enough for one enthusiastic bump.

In another encounter, again in the IFR environment,
but in VFR conditions, the controller was vectoring me
to pass behind a jetliner at the same altitude. When I
reported the jet in sight, he asked that I maintain visual
contact and pass as closely as possible behind it. I passed
about a mile behind it, and even though we were at the
same altitude, there was quite a bump. This was the
airplane with the G-meter on board, and it showed how
certain atmospheric conditions can preclude or at least
slow the settling of vortexes. We were flying right at the
top of the thermal turbulence of summertime, and the air
had a very buoyant feeling. There was enough lifting to
keep the vortexes from settling but not enough turbulence to destroy them in the 20 or 30 seconds that passed
after the jet went by and before I passed through his trail.

Wake turbulence has been a factor in a number of serious accidents at airports, where light airplanes have been flipped by a vortex right after takeoff or before landing. It has also been implicated in at least one in-flight breakup. Any time there is a larger or heavier airplane around, plot against its vortexes.

Jet Blast Avoidance

The blast of the jet engines is often a problem at big airports. Taxiing a light airplane behind a jet, especially one that is applying power to start moving, can result in a topsy-turvy trip across the ramp. Even idle power can get you, if close enough. I taxied close behind a 727 at Washington National one rainy day after being assured it was at idle power, and my Skylane must have been shoved 20 feet sideways when it passed behind the thrust line of the jet engines. Fortunately the ramp was wet, and the airplane just slid. If it had been dry, I'm sure a wing-tip would have been wiped out. I don't taxi close behind them now, and I ask ground control if the jet is at idle when within a few hundred feet of one's blast.

Jet blast can be a hazard to landings at a large airport if a jet happens to have its tail pointed at a runway. It could produce a strong crosswind, and if a jet on the ground has its rear pointed at your runway from reasonably close proximity, beware.

As aviation progresses, airports are probably going to become more and more the exclusive domain of one segment of aviation. General aviation may never be excluded from some of the biggest and busiest airports, but our desire to go to those places may just fade away. I know that in my travels I have come to avoid some big airports like the plague. Atlanta Hartsfield, for example, is one of the most unpleasant airports in the nation to use in a light airplane. Taxi distances are great and you wind up in a long line of taxiing jets, concerned about the jet blast from the one ahead every time it moves. The hazards of wake turbulence are certainly there, and delays are frequent. Many airline passengers feel the same way about Atlanta, but they don't have a choice. The general aviation pilot does and can land at one of the excellent general aviation airports in the area. Before long we'll avoid almost all the terminal airports as a matter of choice. Until then a little planning avoids problems as we fly light airplanes at busy airports.

10
Night Flying

A good friend of mine arrived about an hour after dark, following a long cross-country flight. He was late getting in—arrival was planned for just before sunset—and I was a little concerned as I waited because he had once remarked that he had never flown alone at night.

As I was helping him get his gear out of the airplane, he said that this was only his second night landing. His first had been with an experienced pilot in the right seat. He hadn't had any formal night checkout, but said that it was easier than he thought it would be. Then came contradictory war stories about having trouble finding the airport and being too high and almost overshooting.

To me, night flying is much more difficult than daytime flying. Everything in the cockpit is obscured by darkness and even the best lighting is nothing like daylight. Navigation is more demanding, simple mistakes can be more serious, and making correct interpretations of the distance and movement of other traffic can be quite a challenge.

The simple business of adequate vision in the cockpit is important. One day I was flying a business jet simulator in which it's always nighttime. The cockpit lighting is the same as in the airplane, but in adjusting the lights I never did strike the happy balance between bright enough to see everything and low enough to preserve my night vision. When the lights were bright enough to read the labels on all the switches, there was too much ambient light in the cockpit. So I turned them down, donned reading glasses, and squinted to read instrument approach plates and labels for switches and circuit breakers as I worked with emergency and IFR procedures. A while later, I flew the real airplane in the daytime, and just having good light in the cockpit made it easier to fly than the simulator.

It's obvious that you should fly at night only in airplanes with which you are *very* familiar. Most of us probably couldn't pass a blindfold test in our airplane—in which you show that you can find and operate anything and everything in the cockpit without looking at it—but for night flying we should be able to do so.

I "learned" night flying in a Piper Super Cruiser, in the early fifties. It was a simple airplane, with no radio and only basic instruments. There was no landing light, only the green one at the right tip, the red one at the left, and the white light at the tail. There was no panel lighting at all. A flashlight had to suffice.

The blindfold test was easy in the Super Cruiser.

There were no switches to tend. The only things requiring my touch were the throttle, the control stick, the fuel valve, and occasionally the carburetor heat.

I flew this simple airplane many a mile at night, navigating by pilotage. The 460-nautical-mile route most often flown (between Moultrie, Georgia, and Camden, Arkansas) was a familiar one, and navigation was simply a matter of identifying the towns as they moved by in the darkness below. With no compass light I just flew from town to town—shining the flashlight on the compass would wreck my night vision. All I allowed was an occasional look at the chart. These night flights could be safely conducted only in excellent weather; when over one town, I had to see the next in order to make this form of navigation work. I honestly can't remember how I maintained altitude, but it must have been a combination of power setting and pitch attitude, maintained visually, and an occasional flash of light on the instrument panel.

Flying pitch attitude visually in a dark cockpit was relatively easy, as I could see the horizon quite well on a clear moonless night. Flying with the moon shining was almost like flying in the daytime.

One night the weather for my flight was worse than forecast. Everything started off fine, but as I moved westward through the darkness north of Meridian, Mississippi, it became increasingly difficult to see the horizon. The visibility was lower than forecast, and I could just barely make out the towns. Then I flew over one and proceeded, even though I couldn't see the next. I was tense. After two or three minutes with no town ahead I was doubtful, and in a few more seconds I turned left toward the glow of lights that was Meridian. I had learned that you can't see weather at night, and that unless the night is clear, you can't fly visually in the true sense of the word.

After landing, I went into the flight service station (then called the interstate airways communications station) to see what had happened to the weather. There had been a subtle change. Instead of 25,000 scattered and 15 miles visibility it was 25,000 overcast and 5 miles visibility. There were several forest fires around, and in some areas the visibility had been as low as two or three miles. I must have flown through a smoke area without even seeing it.

Weather

Most of our night training emphasizes landings, but most of the serious problems in general aviation VFR night flying develop en route, flying cross-country. Most of these are weather related—it is difficult or impossible to see weather coming at night. Simple clouds can be invisible, and while thunderstorms are visible because of lightning, they can be tougher to negotiate at night than in the daytime.

An airline crew's encounter with a line of thunderstorms at night is worth relating: The jetliner was moving along on a clear moonlit night toward trouble ahead. A line of thunderstorms was reported between the aircraft and its destination, and the captain elected to fly at 5,000 feet while negotiating the storms.

Concern over severe weather had many people living in the area out watching the clouds, and a good number of them saw the aircraft. All who gave statements reported that the aircraft never reached what they thought was the main line of storm clouds and that there was no cloud-to-ground lightning near the aircraft at the time it broke up in flight. Some said the aircraft entered the clouds before suffering a structural failure; others, farther away from the storm and the airplane, said it was above the low clouds and flying in the clear when they observed the fire that accompanied the breakup. Several witnesses described the weather ahead of the squall line as clear to partly cloudy until a shelf of clouds preceding the storms moved over the area. The base of this overcast was estimated at 1,000 to 2,000 feet. Some witnesses said that the leading edge of the shelf seemed to be rolling forward from the top toward the ground. A wind shift from light southerly to northwesterly, with velocities as high as 60 miles per hour, was reported. Some rain was reported after the passage of the roll cloud, but no heavy rain was reported for 45 minutes. There appeared to be a U shape in the line of clouds. Two funnel clouds were reported in the area a few minutes after the jetliner broke up in flight.

Because of the clear conditions preceding the line of weather, the crew could see the weather from the flight deck. Eight minutes before the accident there was cockpit discussion of a hole in the line and a request for a deviation left of course. The center controller reported that the line appeared pretty solid, and there was further cockpit conversation about deviating. This conversation ended with: ". . . we're not that far away from it . . ." The last intelligible transmission on the cockpit voice recorder tape was "ease power back . . ."

Clearly, this airplane flew into an area of severe turbulence—if not tornadic activity. It's pure speculation, but the crew may have judged that the appearance of clouds in the area of penetration, plus the radar picture, and perhaps the lack of cloud-to-ground lightning, indicated a satisfactory path. If it had been daylight, they would have been able to make a much better visual assessment of the churning low-level roll cloud—formed as it was as the cold air from aloft rushed down out of the mature thunderstorm cell and clashed with warm and moist air being sucked into the storm from ahead. This must have been more apparent to those viewing the clouds from the ground than it was to the crew moving toward the storms at high speed.

When learning to fly in the late twenties, my father asked an instructor what one thing would help most in flying. The answer was "To be able to see the air." In the case of the airliner and the thunderstorm, the crew could have seen at least some of the air had it been daylight. The roll cloud would have been clearly visible and would have warned them away from the area.

Another accident, this one in a general aviation airplane, shows how the differences between a rock and a hard place can become acute at night. About an hour and a half into a VFR cross-country flight the pilot realized that his groundspeed was lower than expected. A check of weather showed that the weather at a planned refueling stop was deteriorating, and the destination airport was changed accordingly.

About 50 miles farther along the way, the pilot lost visual contact with the ground. Then the pilot entered instrument flight conditions. After passing a VOR, the pilot said he was unable to stay on the desired inbound course to the next VOR. The pilot attempted to assure himself of his position; however, he later stated that he did not believe the indications he was receiving from the onboard navigational equipment and he became confused.

His next step was a call to the air route traffic control center and a request for assistance. Through use of the transponder, with verification from the aircraft's navigational equipment, radar contact was established. The pilot stated that there was an hour's fuel on board, and he told the controller that he'd like to get down.

There was discussion of weather, and another destination was selected. Radar vectors were furnished to the flight, which was now being handled as an IFR operation. The pilot was instrument qualified, and the controller asked him if he could make an approach using a nondirectional radio beacon if the controller read the approach procedure to him. The pilot said he thought he could.

There followed some confusion over whether or not the pilot had the airport in sight. The pilot said he had the "beacon," and from the transcript of their conversation it appears that the controller thought he meant the rotating beacon at the airport. The pilot actually meant he was receiving the nondirectional radio beacon on his ADF receiver.

Radio contact was lost as the aircraft descended. It was in heavy rain and the pilot could not see the ground from 3,000 feet, but at 2,500 feet he could see lights glowing. He then descended to 2,000 feet, where he could see the tops of houses and other objects but could not see the airport. He made a 360-degree turn in an unsuccessful attempt to find the airport. The pilot then started climbing, and as soon as he was high enough to contact the controller, he reported missing the approach.

The controller gave the flight a vector heading to another airport and asked for a report on the amount of fuel remaining. The pilot replied that he had 30 minutes' fuel on board.

A few minutes later the controller told the pilot that the rain showers seemed to have passed the airport that he just missed and asked if the pilot would like to return to that airport. Because of winds and the amount of fuel reported by the pilot, the controller wondered aloud if he might be close on fuel getting to that next airport.

The pilot declined to go back to try the first airport again and soon was in better weather conditions and had the rotating beacon at the next airport in sight. He was close to the airport when the engine failed. He selected an open field and landed successfully, but struck a row of trees in the landing roll. The cause of the engine failure was fuel exhaustion.

There were other factors not covered in this brief synopsis of the accident. The weather was worse than forecast, the winds aloft were strong, and the pilot's descent for the first approach was gradual and time-consuming because of turbulence. On relating it to night flying, though, it is reasonable to question whether this accident would have happened if the same conditions had existed in the daytime.

The inability to see clouds before flying into them at night got things off to a bad start, and the difficulty of identifying things on the ground in low visibility conditions and darkness made the pilot's search for an airport fruitless.

The additional demands brought on by poor illumination in the cockpit might have been a factor. The National Transportation Safety Board report on this accident cited the fact that the pilot did not use the approach chart available to him in making the first approach. Instead, the controller read the approach to the instrument-rated pilot. The NTSB also noted that a flight log was not used.

Too many pilots tend to neglect paperwork at night simply because it is more difficult. I know that I keep a good flight log in the daytime but tend to do only what is necessary at night. Marking down the takeoff time is a simple matter by day; at night it means getting some illumination on the writing surface, perhaps at a time when flying a combination of visual and instruments, and when both a distraction and light in the cockpit are undesirable. That's no excuse for omitting necessary paperwork, but it illustrates why we often don't do things as precisely at night.

Cockpit Lighting

When reading charts and doing paperwork, it's often difficult to decide on the proper means of illumination. Airplanes usually have overhead lights and map lights, but often these seem too bright or tend to shine in the wrong place. The map light on my Cardinal RG was beneath the control wheel. It was fine for illuminating my lap, but it didn't do much for a chart and was impossible to use when posting a flight log.

A small flashlight is usually the best solution, and this means having two flashlights on board—one that's bright enough to use in preflighting the airplane is too bright in the cockpit.

Besides using it to read charts, the flashlight is mandatory if the cockpit lights go out. A friend of mine likes to tell his war story of what happened when the lights went out, and he invariably tells it when you start complaining about the stench from his cigar.

You wouldn't expect this opening line to come from a physician: "Cigars are good for you, in fact one saved my life one night."

He was tooling along through a dark night when, zap, the electrical system of the airplane failed and the instrument panel was plunged into darkness. No flashlight. What to do? Light a cigar, what else?

This pilot claims that when he was puffing away at his cigar the glow at the tip enabled him to read the instruments, maintain a heading, consult the chart, and locate an airport. Despite the heroics of the cigar, I bet it made the cabin foul. Flashlights are better.

Fuel

Most of us don't fly often at night. As a result, we aren't aware of the limited activity of the night aviation community. While there are usually plenty of airports for refueling by day, there are precious few at night. This is probably one reason why there are a lot of fuel exhaustion accidents at night. If a landing for fuel is made and the airport turns out to be unattended, a pilot might well be stranded in the middle of nowhere without fuel to fly anywhere. The serious trouble comes when the pilot

skips a stop because of that possibility or tries to stretch
gas to the next airport after finding one deserted. The
solution is in calling ahead from the last daylight stop. A
few bucks for a phone call is inexpensive insurance that
someone will be waiting to pump gas later in the evening.

Departures and Arrivals

While most serious night problems are related to
cross-country flying and weather, that doesn't mean that
departures and arrivals don't take their toll.

Starting with takeoff, should a pilot use taxi and landing lights on the takeoff run? Some do, some don't. Taking off without lights avoids the transition from a brightly
lighted outside reference—the runway illuminated by
your lights—to flying without any lighted outside reference. The initial stages of a night departure are often as
pure a form of instrument flying as found in cloud. Even
if you leave the lights off, flying off a brightly lit runway
into darkness is a dramatic transition, and anything that
makes it easier is worthwhile. On the other hand, using
the lights on takeoff probably makes directional control
a bit easier, and if there is anything on the runway, they
might illuminate it in time to make avoidance possible.

Using taxi and landing lights on takeoff assures you
that approaching traffic will have a better chance of seeing you—especially important at an uncontrolled airport. There have been collisions at uncontrolled airports
involving aircraft landing in the direction opposite departing aircraft, or aircraft using different runways. It is

likely that if the airplanes involved had been illuminated to the maximum, "see and be seen" would have saved the day.

Instruments

The transition to instruments after a night takeoff is challenging. At New Orleans' Lakefront Airport, where north departures leave the airport over the ink-black darkness of Lake Pontchartrain, there have been numerous cases of airplanes flying into the water. It's an instrument departure and must be treated as such.

On any night takeoff, the first thousand feet of climb is critical. What's out there? Is it a brightly lighted area, or is it sparsely populated? Brightly lighted, there will be visual cues. Sparsely populated, there will be none. Even if it's brightly lighted, the visual references in climb might not be too valid.

I start night takeoffs with a thorough search of the sky for other traffic *before* beginning to roll. Then I'm as certain as possible that there won't be any traffic conflict in the initial climb. I make night takeoffs in VFR conditions with landing and taxi lights on. If IFR, leave them off to avoid flying into the overcast with the lights on.

For night takeoff, I run to a speed five or ten knots above the normal rotation speed, then lift off. When a positive rate of climb is indicated, the altimeter has started up, and the airspeed is healthy, I retract the gear if the airplane is a retractable. Pitch attitude is flown, with airspeed the judge of the proper pitch attitude. The target is the best angle-of-climb speed unless this results in an excessively high deck angle. In this case, the best rate-of-climb speed is used. I hold this initial speed at full power until 1,000 feet above the ground. The flying is done by reference to instruments. If it's VFR, I look outside only to check for traffic. At 1,000 feet, I transition to normal climb attitude, flaps are retracted, power is reduced, and if VFR, the search for traffic is emphasized. Landing and taxi lights are extinguished after leaving the airport area. It's identical to a daytime VFR departure except for the reliance on instrument flying technique in the first 1,000 feet of climb.

In daytime VFR we climb rapidly for the first 1,000 feet for two primary reasons: (1) to achieve as much altitude as possible as quickly as possible for maximum flexibility in case of an engine problem and (2) for noise abatement. Altitude is the best means of quieting the airplane for those on the ground. There's a third reason at night: to avoid the possibility of flying back into the ground or into obstructions. During daylight we can see outside. That's not true at night.

Approaches and Landings

On a night departure, we do the best we can to avoid obstructions by climbing as rapidly as possible. Too often, we don't do the logical thing on arrival and descend as rapidly as possible from a known safe altitude to the runway. Unfortunately, airplanes hit trees short of the runway and fly into high terrain around airports with some regularity. Remember, while man-made obstruc-

tions are required to be lighted, obstructions put there
by the Lord are usually only charted. Avoiding these is
a matter of keeping up with position in relation to terrain
and flying at altitudes that give a safe margin above the
hard stuff.

A good way to avoid rocks and trees is to use IFR
procedures even in VFR conditions. A confirmed VFR
pilot might think this an excessive obsession with IFR,
but see how useful it can be.

Take Hot Springs, Arkansas, for example. Everybody
knows that Arkansas is flat country, right? Wrong. Only
part of Arkansas is flat. The north central, northwest,
and western parts of the state are hilly, and a lot of pilots
have rammed those hills because they didn't know they
were there.

Approaching Hot Springs on a dark night after the
tower has closed, look at the sectional chart. There are
some little symbols north of the airport indicating obstructions. The most ominous is 1,518 feet above sea
level, 253 feet above the ground. That's not very high
above the ground. There are some terrain contours
around, but they don't appear significant and might not
appear at all under red or poor cockpit lighting. Fly on,
nothing big sticking up there. Get closer, though, and
you might suddenly realize that there is a 1,000-foot
difference between airport elevation and that obstruction just north of the airport. The shadowy figures out
there to the north and northeast of the airport are hills.
The obstruction is on a *hilltop*. How high is high enough
to bother? The sectional is both hard to read and sometimes deceiving.

Go now to the Jeppesen instrument approach plate

and read "CAUTION: steeply rising terrain north and northeast of airport." That tells you something important. So does the note on the minimums for circling to land in IFR conditions. Circling is *not authorized* north of the airport between runways 4/23 and 13/31. You sure don't get that off a visual chart, which unfortunately is all most pilots carry when flying VFR at night. More study of the Jeppesen chart shows that you can circle at 1,220 feet above sea level south of the airport within one mile, or at 1,400 feet within two miles. The airport elevation is 535, so you know that a normal pattern can be flown on that side of the field. For maneuvering in the area, 3,100 feet is the minimum safe altitude within 25 miles —except to the southeast, where it's 2,600 feet. That's also from the approach plate.

You can get information like that from several other sources, but the approach chart has it all in one place. It's easy to read, with safe altitudes precomputed for you.

Better Safe Than Sorry

Fayetteville is another example in Arkansas. The terrain doesn't look bad on the chart—just some contour squiggles, an obstruction 1,982 feet above sea level to the northeast, and an obstruction 2,349 feet above sea level a few miles east. The 2,349-foot one is 469 feet above the ground, but it's also over 1,000 feet above the 1,251-foot airport elevation—obvious when you study a chart at home by the fire, but not so obvious on a bouncy night flight.

Go now to the Jepp chart and look at the airport diagram. The place is obviously surrounded by hills, close to the airport. Look at the circle-to-land minimums for advice—2,040 feet is the minimum altitude within a mile of the airport. Come over the VOR at 2,300 feet, descend to 2,040 (if flying a light airplane) on the 138-degree radial until within a mile of the airport, then fly the pattern and land. Sure beats guessing. In the daytime, good terrain clearance procedures are obvious. That's not so at night. These numbers are examples—never use anything but a current instrument approach chart for guidance.

Here's how unfamiliarity with airport surroundings can cause problems when flying to Fayetteville. The fact that it was a daytime flight might make it seem unrelated to this discussion of night flying, but it is applicable.

We took off early in the morning for an afternoon football game. The flight from Little Rock was interesting. There was some ice along the way and I had to climb to 10,000 feet to get above it. When the time came to descend, I waited as long as possible and then descended rapidly.

Fayetteville weather was 1,000 overcast, five miles visibility, and a 30-knot northwest wind laying a 30-degree crosswind on the 6,000-foot runway. I planned to use the VOR approach, with a circle to land on runway 34. I foresaw that it would be turbulent down low over the rough terrain and warned my passengers that the circling approach in our light airplane might be uncomfortable. I was number one for the approach. Number two was a chartered DC-9 carrying the football team opposing the University of Arkansas. The airline was one that didn't usually operate in the area, and there was some weather-related discussion between the crew and the controller. Finally, the crew decided that it had minimums for the circling approach, that the crosswind wasn't beyond limits, and that the runway was long enough for landing at the weight they were flying. From the length of time it took to hash this out, I got the impression that everything was just barely in order. The jet started its descent, staying 1,000 feet above me.

There was then a call to the airliner, same frequency but a different controller's voice.

"Ah, say, are you guys familiar with the airport at Fayetteville?"

The crew replied that they were not.

"Well, I tell you what: it's surrounded by hills, it's nasty turbulent down there when the wind blows, you'll have a tailwind on base, the ceiling right over the airport often is better than it is in the pattern out over the hills, and well, I just sure wouldn't want to be on a DC-9 trying to land there today."

There was a silence. The airline crew then made a wise decision and headed for Fort Smith, to put the team on a bus to Fayetteville. Another football team charter flight flying a nonprecision approach in minimum conditions to an unfamiliar airport in hilly terrain didn't do so well that very night: it hit a low ridge before reaching the airport and all were lost.

It's a well established fact that the night accident rate is far worse than the daytime rate. It's no fluke. There's good reason for it. Even though night flying might seem to be almost as easy as daytime flying at first encounter, there's more to it. Perhaps the important difference is not in basic flying technique—the wheel and the pedals can't distinguish between daylight and dark—but in the general characteristics of the night environment. Everything is a little more difficult and the consequences of errors are markedly more serious. The visual cues that keep us out of trouble in the daytime are gone, and like it or not, night flying is best done IFR.

11

Mountain Flying

A primary point in the relationship between terrain and airplane is that mountains are where you find them, not necessarily where you remember them as being. As mentioned in the preceding chapter, those hills in Arkansas are struck by airplanes with tragic regularity, often because the pilot didn't really think there would be mountains in Arkansas. Over one holiday period a few years ago there were three airplanes missing at one time in the hilly part of that state. They were all lost while trying to fly VFR through a stationary weather system that was spreading low ceilings and visibility over the hills; it took days of searching to find all three. There were no survivors.

An east/west collection of ridges south of Fort Smith, Arkansas, punctuates a north/south VFR flyway and has been hit by general aviation, military, and airline aircraft over the years. It didn't get any one of the three airplanes lost in the holiday period mentioned, but the toll has been so high in the area that there is now a warning printed on charts.

The mountains of the east, especially the Alleghenies in Pennsylvania, claim their share of airplanes, too. Headed west from the Atlantic coast, pilots flying VFR hit the first ridge with some regularity, surprised as they fly from over flat country into that first obstruction.

The big mountains are in the west, though, and for someone used to flying east of Denver there's a special challenge to the really high country. The first trip into or through the area is something to be approached with care; you might be a multi-thousand-hour whiz over flat country, but you'll start out as a zero-time mountain man or woman.

My first Rocky Mountain experience was in May 1962, while ferrying a new Cherokee from Vero Beach, Florida, to Portland, Oregon. My mountain checkout consisted of someone telling me to fly only with good visibility, to stay on the ground if the wind at the ridge level was forecast in excess of 30 knots, and not to get in a hurry.

As I moved toward the mountains, I was enthusiastic about the tailwind that had developed. It was really bad news, though, because wind flowing up the slopes (called an upslope condition) makes clouds, just as any lifting action can cause condensation and cloud formation.

Because of weather, I kept moving northward, to put off the time of entering the mountainous area. One of the lads in Portland later kidded me about this. He said that everyone thinks they can skirt the mountains, and they expect someone will eventually fly all the way to the North Pole before turning left.

The moment of truth came at Sheridan, Wyoming. If I was to get to Portland, I had to start into the mountains. The question was whether it would be better to hole up at Sheridan for the rest of the day or to fly some more. I sought the advice of a local pilot, who, after studying the situation, said that it should be okay to go at least as far as Billings, Montana. The reports looked fine for this —Sheridan had 5,000 scattered and 30 miles, and Billings 5,000 broken and 60 miles—but my advisor reminded me that many mountains to the west of Sheridan would be obscured by clouds. He added that in the mountains you use sequence reports in conjunction with knowledge of the terrain and the effect of circulation in the area. Billings might be fine for local flying with 5,000 broken, but in the mountains it could be impossible.

It was an interesting trip. I actually made it on to Helena, Montana, that day, one mile at a time. The destination changed at least once as I followed the paths of least resistance—wide valleys with a road, railroad, or river to follow. At Helena the weather folded and I had to wait until the next day, when again with the advice of local people, I moved cautiously on.

On that trip I flew through, not over, the mountains. The technique was simple: I flew only with good visibility. The worst it got was six miles in rain, right before I landed at Helena. As long as I could see six miles in front of the nose of the little airplane, I wasn't going to hit anything with it.

Mountain Strips

Some years later I returned to Helena and got Jeff Morrison, the fixed-base operator there, to teach me a bit about flying into some of the mountain strips. It was a beautiful and interesting experience.

We started with a fairly easy one and finished the short course with a runway at the end of a valley. The strip couldn't be seen until we rounded the last bend, and a commitment was made to land before the strip was in sight. The valley was too narrow to turn in, and the airplane would not have outclimbed the terrain at the other end of the strip. It is obviously imperative to get dual from a talented native before trying any mountain strips.

In explaining why he didn't use turbocharged airplanes in his operation, Jeff Morrison made a strong case for valley-flying. Fly the valleys and *know* exactly where you are at all times. Another person suggested that the engineers who figured out the railroad and highway systems were good at finding the best terrain for building, as well as the lowest route between two points, so following highways and railroad tracks is a good bet. In fact, those are the primary navigational aids for light-airplane flying in many mountainous areas. The airways can't bend to follow the lowest terrain, so they become of limited use unless you are flying a turbocharged airplane with pressurization or a full oxygen bottle.

Light-airplane IFR in the mountains can be a very bad deal without turbocharging. The minimum en route altitudes are often near the ceiling of normally aspirated airplanes, which means a drastic reduction in options.

With the airplane operating at its ceiling and with jagged rocks as close as 2,000 feet below, there's no margin for icing, downdrafts, or anything else. You might say that IFR in the Rockies without turbocharging is optionless flying, which is not so good.

Even with turbocharging, the mountains can be formidable IFR obstacles, as I found one day when westbound in a Turbo Centurion. Wind is very definitely weather in the mountains. The forecast for this day was for 40 knots at ridge level; I felt I could handle that with turbocharging. My plan was to go over the big ridges west of Pueblo, Colorado, and then on into Gunnison, Colorado. I thought it wouldn't be too bad around Gunnison because there aren't any big ridges immediately to the southwest, the direction from which the wind was blowing.

I went up pretty high, to Flight Level 240 (24,000 feet) to try to stay in smooth air, but as I got closer to the mountains there were up and down drafts and the air became more and more turbulent. It was not a real operational problem, but when I looked around and surveyed my four passengers hiding behind oxygen masks, I could discern light shades of green beginning to show around the edges. Barfing with an oxygen mask on would be the ultimate bad scene, so I gave up this day and stopped for the evening in Colorado Springs. The turbocharger would have gotten us across safely, but not comfortably.

Another trip from Pueblo to Gunnison in a Turbo Centurion was reflective of clouds as weather. We were comfortably on top at Flight Level 220 approaching the mountains, but the effect of the wind flowing over high ridge west of Pueblo did its thing to cloud tops, and we were soon in the stuff. Even a climb to Flight Level 240 didn't help. Ice started forming on the airplane, and if we had had a long distance to go before a lower altitude would be available, it would have become a problem. As it was, the descent into Gunnison was started before too much ice accumulated.

Engine Failure

We had two Turbo Centurions on this photo mission and there were some interesting moments on a later leg out of Gunnison. The other airplane was a few miles ahead of me and 2,000 feet above, at FL 240. We were droning along IFR, in clouds, no ice, and everything seemed rosy. Then the pilot in the other airplane made a chilling announcement: "Mayday, my engine has quit."

At first his options seemed slim. We were on a 14,000-foot MEA airway, which meant there was some high stuff along the way. He was headed down toward it. For some reason I don't think I was at all concerned about separation even though he was in front of me and descending through my altitude. Instead, I was trying to solve his problem with body English when he announced the solution. The airport at Montrose, Colorado, was actually within gliding distance, and he said he'd be able to maintain terrain clearance altitudes on a powerless trip to the airport. Later he announced that he had some power back, and the landing at Montrose was uneventful. The problem was a blown turbocharger hose. The engine was running, but the turbocharger was out of service. At 24,000 feet he thought it was all gone, but down in thicker air there was power to use.

The lesson to me was that there are options even in

the mountains. Good thinking can salvage potentially bad situations.

Wind and Clouds

Wind demands as much thought as clouds when flying in the mountains, and even when abiding by the 30-knot maximum at ridge level, you can have some bouncy up and downs. Downdrafts can be strong with 30 knots of wind, and turbulence can bang you around terribly. And don't let light surface winds fool you. The wind at ridge level can be much stronger.

The worst possible deal is to have the wind at an angle to ridge orientation when you are over or downwind of the ridge. The wind velocity can increase as the flow moves through passes and notches in the ridge, and as it spills over the ridge it can create severe turbulence. In extreme situations a rotor can form downwind of the ridge.

A rotor cloud looks a lot like a plain old cumulus, but it is in complete circulation. If you ever flew into one you'd probably call it a horizontal tornado. The top of the rotor cloud won't be a lot higher than ridge level, but updrafts and downdrafts (and turbulence) can extend up higher. A good rule is to fly 50 percent higher than ridge height above the ground to avoid rotor turbulence. For example, if the ridge level is 12,000 feet and the general terrain is 5,000 feet, making a difference of 7,000 feet, the rule of thumb would suggest flight at least 3,500 feet above the ridge, or at 15,500 feet or above. This would in no way eliminate downdrafts or turbulence, but it should keep the airplane clear of the rotor.

Lenticular clouds are pretty little lens shapes that signal what's called a mountain wave. Soaring pilots like this because it means the presence of good and easily identified updraft conditions. Power pilots had best have the ship rigged for rough running if there are lenticulars, especially if flight is to be in the airspace between ridge level and the lenticulars.

Lenticulars form at the high point of the undulating wavelike flow downwind of the mountains. This high point is usually found where there is a temperature inversion, or at the tropopause (the point where the decrease in temperature with elevation ceases).

What about going below the ridge height when lenticulars are present? Downwind of the mountains I'd prefer to avoid that until quite a lot of miles away from the ridges. And always consider the attendant downdrafts.

One of the key technique items in mountain flying is in learning to visualize what the wind is doing to the air. For example, if approaching a ridge and flying with a tailwind, you'll know that there will first be an updraft, which will turn into a downdraft as the ridge is passed. Flying conditions will likely be smoother approaching the ridge than they'll be after passing it, when the air is spilling over the ridge. Relate it to water flowing over rocks. The water is smooth until it reaches the rocks; then it is disturbed for a distance downstream after passing over the rocks. Flying toward a ridge with a tailwind gives the advantage of having the updraft before reaching the ridge, but it also offers the promise of the worst turbulence after passing the ridge.

Going the other way, toward a ridge with a headwind, the item of technique is to have a way out if the downdraft overpowers the climb ability of the airplane. Real mountain flyers approach ridges at a 45-degree angle, so they can more easily turn away if the downdraft is too bad. And they go at the ridge with enough altitude to lose some and still make it across comfortably. If the wind is forecast anywhere near the 30-knot limit, I don't think that I'd like to go toward a ridge with less than a 2,000-foot advantage.

You can't really "see" the air when flying in the mountains, but by studying the situation carefully you sure can avoid a lot of surprises.

Stall/Spin

Stall/spin accidents are common in the mountains. These can be related to wind, blind canyons, and to a combination of high density altitude and poor pilot technique.

In any case, the loss of the familiar horizon reference can make visual attitude flying (thus airspeed control) difficult for pilots used to flying in flat country. When looking at a mountain from a distance, the horizon is at the base of the mountain, not at the top. With the nose pointed below the ridge line, you might think it down when actually it's a bit up. When close to a mountain, as right before crossing a ridge, the horizon is hard to identify. If a downdraft should drop you below the ridge level ahead, the horizon would be out there somewhere, but the instruments on the panel have to be allowed to tell the story. Trouble can come fast when a pilot lets the airspeed decay and then tries to turn steeply away from the ridge or turn around in a narrow valley or blind canyon.

Flying into a blind canyon is a product of poor navigation. Again, roads and railroads are the best things to follow, but do be aware that following a road or railroad doesn't preclude flying into a blind canyon. There are tunnels, and airplanes don't fit into these very well. The chart will show tunnels, and if the route is carefully surveyed before flight, you'll know the location of such things.

Thin Air

Density altitude is very directly related to stall/spin and blind-canyon problems.

The service ceiling printed in our pilot's operating handbook is density altitude—a Skyhawk will operate at 14,200 feet density altitude, but that is not 14,200 feet on the altimeter. It'll go to a higher indicated altitude on cold days and a lower indicated altitude on warm days. If it's +25 C at 8,000 feet, the air is thin and the density altitude is up around 11,000 feet. Even though the altimeter is on 8,000, the airplane's climb is referenced to the 11,000-foot density altitude, which is to say that it won't climb very well. I remember thinking very bad thoughts about the climb ability of my Cherokee Six one day when it was actually performing exactly as the book said it would perform. Trouble was, I was looking at the altimeter when considering its rate of climb when I should have been looking at the altimeter *and* the thermometer.

Headed toward a ridge, close to the airplane's ceiling, in turbulence, trying to climb a bit, the airspeed low —that's quite a setup for an inadvertent stall and possible spin.

High density altitude also plays hob with takeoff distances and, to a lesser extent, with landing distances. Careful preflight calculations and a conservative determination of required runway length is what it takes to avoid problems.

As discussed earlier (page 19), it is suggested that the minimum required runway length be 2.2 times the ground roll shown in the pilot's operating handbook. For a Cardinal RG operating at gross weight, +30 C, and 6,000 feet elevation, that means a minimum runway length of 3,828 feet. Even with that conservative approach, the airplane will be only about 50 feet high over the end of the runway on departure, the climb rate will be but about 500 feet per minute, and the climb gradient but about 375 feet per mile. The message: In the mountains you had best go past calculating runway length and relate climb gradient to the terrain, leaving a comfortable margin that allows for the effect of turbulence and downdrafts. Without this margin, the ingredients for a stall/spin or collision with terrain are there, in spades. The only safe operation might be at a weight lighter than gross, or in the cooler temperatures of early morning or later afternoon. In hot weather and high country, four-place airplanes usually become two-place airplanes; some are marginal at that. If your machine has six seats, four or less are probably usable at high density altitudes. Turbocharging helps, but it doesn't repeal the laws of gravity and aerodynamics.

The solution to the high-country stall/spin problem is in recognizing that the airplane is not going to perform well and in keeping it flying at a safe airspeed and away from situations where there will be a strong temptation to try to make the airplane do something it is incapable of doing.

Solitude

When mountain flying, I always have a strong feeling of being alone. There are few people on the ground, and radio chatter is sparse or nonexistent. The luxury of going IFR and always having a good airport with an instrument approach nearby is missing. Unless in a high-flying airplane, it's VFR, you, and the terrain. The stakes can't be any higher than in other forms of flying—the ultimate reward of ineptness is always the same—but in the mountains, especially the Rockies, the airplane is even less forgiving than elsewhere. If you do something that isn't bright, like launching when the wind is howling or flying IFR when there's ice and no way to get above it, the inevitability of sinking in over your head seems more assured than in flat country. It can be a great distance to a safe haven. Or if the airplane is icing, the possibility of settling out of the bottom of the clouds into relatively friendly air over open spaces is much less likely over the mountains than over the plains. Over flat country inattention to position or a navigational error can result in a temporary period of misplacement; in mountainous country it can result in a trip up a blind canyon if VFR, or flight into obscured terrain if IFR.

Bad Night at Salt Lake

There was a good example of mountainous revenge north of Salt Lake City one night. The airline flight was cleared to descend to 6,000 feet and was told that the

ceiling at Salt Lake City was 1,700 feet. (The field elevation at Salt Lake City is 4,227 feet.)

The first hint of a problem on this flight was when a
crewmember told the air traffic controller that "we're
working with radio problems." They later said, "Okay,
we got . . . a few little problems here, we're trying to
check our gear and stuff right now." Then, after being
cleared to land, a member of the flight crew said that they
wouldn't be landing because of the need to check the
landing gear. They requested to be taken back out about
20 miles, and to maintain 6,000 feet. Then: "Ah, tower,
we're gonna have to, ah, nuts, just a second." Next: "You
put us in a holding pattern at 6,000 on the VOR for a
while?"

The controller cleared the flight to hold north of the
VOR at 6,000 feet, and the flight switched over to a
company frequency to talk with maintenance personnel.
Apparently only one radio was operative, and the flight
remained on the maintenance frequency for about seven
minutes. It flew one and a half holding patterns, and
when the flight returned to the air traffic controller it
had, for some reason, rolled out on an easterly heading,
still level at 6,000 feet. High terrain was directly ahead.

Flight: Oh, ah, hello, ah, Salt Lake, United 2860,
we're back.

Controller: United 2860, you're too close to terrain
on the right side for a turn back to the VOR, make a left
turn back to the VOR.

Flight: Okay.

Controller: United 2860, do you have light contact
with the ground?

Flight: Negative.

Controller: Okay, climb immediately to maintain 8,-
000.

Controller: United 2860, climb immediately, maintain 8,000.

Flight: United 2860 is out of six for eight.

Eighteen seconds later the aircraft flew into terrain at the 7,200-foot level. It was in a wings-level climb attitude at impact.

The National Transportation Safety Board determined that this accident was probably caused by the controller's issuance, and the flightcrew's acceptance, of an incomplete and ambiguous holding clearance in combination with the flightcrew's failure to adhere to prescribed impairment-of-communications procedures and prescribed holding procedures. Contributing was the failure of the aircraft's number-one electrical system for unknown reasons.

The lesson is to know position in relation to terrain. We shouldn't lean on a radar controller to keep us out of trouble, either. There have been cases of misidentification of targets, and of IFR aircraft being allowed to fly into terrain, so it's important to keep your own private running tab of what's going on. And if a mechanical problem commands attention, it should be treated only after a determination is made that the airspace in which the airplane will be operating is free of rocks. This means having visual charts on board for reference or, preferably, staying above published minimums on established IFR routes.

Big Downdraft

Even when IFR on airways there's a requirement to correlate strong winds with the ridges and to anticipate the effects of downdraft activity. An example of what to avoid was found as a light twin moved along IFR, downwind of one of the largest eastern ridge lines. The aircraft was at the minimum en route altitude of 9,000 feet when it experienced a severe downdraft. The pilot reported that altitude couldn't be maintained.

A downdraft doesn't go all the way to the surface—at some point it must fan out over the ground—but a downdraft can take an airplane down to an altitude where recovery is impossible. In this case, the airplane was apparently stalled as the pilot tried to keep from losing too much altitude. This teaches that the minimum en route altitude isn't necessarily a safe altitude when the wind is blowing in the mountains. Even downwind of the relatively low ridges of West Virginia I've seen downdrafts that make it quite difficult to maintain altitude when flying 5,000 feet above ridge level in an airplane capable of climbing 600 feet a minute at that altitude.

Certainly experience is the best teacher in mountain flying. But the primary thing that experience teaches—caution and extra respect for wind and weather in the mountains—is available to anyone, free of charge. Success comes in recognizing that mountain flying is different from and a lot more challenging than flying in flat country, and that the airplane must be allowed to deliver a somewhat lower level of utility when operating in the high country.

12

Moving to Higher Altitudes

Most of us do our initial cross-country flying in airplanes that cruise along quite well at altitudes of up to 10,000 to 12,000 feet. We learn the weather of those altitudes, and we also learn that a person living at sea level usually can't fly at 12,000 feet all day without becoming lethargic and developing a headache. Like the piston engine without a supercharger or turbocharger, we start running out of energy when a couple of miles high.

Turbocharged (and turboprop) airplanes are becoming more common, and more of us are venturing into the higher altitudes. It's normal for turbocharged light airplanes to operate above 12,000 feet, and a lot fly up in the 18,000-to-25,000-foot airspace. This is good because the airspace up there is less crowded than down low, the air traffic control procedures are simpler, and the weather is often (but certainly not always) better, but some pilot requirements go with the high-flying ability of the airplane. You might say that the altitude capability of the airplane should be matched by the pilot.

Instrument Rating Required

To get full value out of a high-flying airplane, a pilot *must be instrument rated.* There's no question about this. It is too easy for clouds to come between a high-flying airplane and the ground. If clouds do intervene, a descent from up high can become a long and dangerous process for a pilot who can't fly instruments.

It's hard to tell VFR from IFR up high, too, as I saw one late-spring day. The flight was from Monroe, Louisiana, to Washington, D.C. Some clouds were forecast along the way, about to the Mississippi River, then it was supposed to be clear on to Washington. I was flying IFR at 19,000 feet (Flight Level 190), and the morning was beautiful after we left the area of heavy clouds around Monroe. But there were some high clouds, and the other pilot on board and I got to talking about how difficult it was to tell whether we'd be operating clear of clouds along our path. The clouds were in thin layers and were reported by ground stations as 25,000 scattered or broken. They were at least 6,000 or 7,000 feet below the reported value and were in relatively thin layers and patches. It was hard to make a determination that we'd be over or under upcoming clouds, and detours around would have required a lot of dodging; even then we might have flown into one. It was clear that while reports were for excellent VFR, and suggested that VFR would be possible at 17,500 feet (the maximum legal VFR altitude in that area at the time), IFR was the only safe way to go.

There have been a lot of real problems with this. Every year, more than one pilot without an instrument rating gets a turbocharged airplane between layers or in a hazy or on-top situation, loses control, and winds up in a long fall.

Oxygen

High-altitude flying carries a strong requirement for an understanding of the need for oxygen. The uninitiated might kid themselves about this, but pilots who have gone through the Air Force altitude chamber sessions know how it works. They have respect for the effects of altitude. (Your local FAA General Aviation District Office should have information on how you can participate in high-altitude training, which the Air Force makes available to civilians.)

When turbocharged twins first came out, a pilot flying one apparently decided that the effects of altitude wouldn't bother him and he set out to climb as high as the airplane would go. I don't know how high he got, but little parts and pieces of airplane fluttered down as a reminder that regardless of what kind of airplane you fly, there's no getting around the fact that oxygen is necessary at higher altitudes.

An important thing to understand about hypoxia—a

deficiency in the amount of oxygen reaching the tissues —is that its effect is not proportional as you ascend. If you can take off, climb to 10,000 feet, and still feel pretty good, this by no means indicates that you can climb on to 20,000 feet and feel about half as well or function half as well. It doesn't work that way.

As pressure decreases with altitude, our ability to use the oxygen in the air remains pretty good up to about 10,000 feet. When you go higher, the ability to use the oxygen in the air starts to drop off rapidly. Our ability to think and act deteriorates. At some point, up around 20,000 feet, there's too little pressure for us to get the oxygen required to function at all. Out we go, cold.

There are two ways around this. One way is to pressurize the cabin to keep the pressure up to what it was at or below 10,000 feet. Some people hold the misconception that oxygen is pumped into a pressurized cabin, but it is not. At the higher altitudes we fly, oxygen is there in plentiful quantities; all we need is for the pressure to stay high enough for the body to be able to use the oxygen. Pressurization is a matter of sealing the cabin and pumping air in.

The second way to handle altitude is to breathe oxygen, taking in much more than is naturally there to make certain the body is getting enough. Wise pilots use 10,-000 feet as the maximum operating altitude without oxygen. If the cabin isn't pressurized, that means using oxygen above 10,000. If the cabin is pressurized, that means not operating at altitudes where the cabin pressure is above 10,000 feet.

There have to be exceptions to the rule, as there are people who live above the 10,000-foot level without puffing on oxygen. Perhaps they develop a tolerance, and perhaps there are other factors involved, but their ability doesn't help us as pilots. To operate an airplane safely we need to be at our best, and that means steering clear of hypoxia by following conservative guidelines.

Age and Condition

Age, general physical condition, and drinking and smoking habits have an effect on altitude tolerance. When I was younger, I could fly at 12,500 feet all day without feeling any effect of high flight. At 46 I feel out of place at about 10,000 feet and have a headache for a day after flying near that level for long. By day I can make flights without using the reading glasses that I must carry per a note on my physical; at night I can't read a chart without glasses—especially when above 5,000 feet. Night vision *is* one of the first things to go as the pressure decreases and we are able to use less of the oxygen in the air. Birthdays help illustrate this very clearly.

How High Is High?

When considering how high you might want to fly *with* oxygen, look at the worst case and contemplate the effects of a sudden oxygen system failure.

If you are flying at 18,000 feet and the oxygen goes off, you might or might not eventually deteriorate to a totally unconscious state. It is said that the time of useful consciousness at that altitude, without supplemental oxygen, is around 30 minutes, plus or minus, depending on your condition. If you did manage to stay conscious, you'd still be a very unreliable pilot after only a few minutes off the oxygen.

One oxygen system that I use has a flow meter in the hose. If the little red slug in the line isn't visible, the oxygen is flowing. The oxygen bottle is usually behind me, and the flow indicator is at my side or behind me. It's not where I'd catch it the instant of failure, but hopefully I'd catch it within a few minutes. I don't feel that an oxygen system failure would be extremely critical at 18,-000 feet.

When you climb higher, to 24,000 feet, which is an approved and easily attainable altitude for many turbo-charged light airplanes, things really change. Even though you've climbed only 6,000 feet, an oxygen failure would be critical. Unconsciousness would come in three or four minutes or even less. That's not much time. The flow stoppage would have to be detected instantly and an emergency descent would have to be instituted immediately. Even at a 2,000-foot-per-minute rate of descent, you'd be in the gray area between 18,000 and 24,000 feet, where unconsciousness could come quickly, for three minutes. And you'd be in low enough pressure to be hypoxic until down at 10,000 feet, which would take seven minutes at 2,000 feet per minute. If the air happened to be turbulent, the descent might have to be at less than 2,000 feet per minute to keep the airspeed at maneuvering speed. Your old brain might barely be hanging in there while you are doing some of the most demanding flying there is.

Even though flight at 18,000 feet and 24,000 feet seems very similar from the standpoint of how things look, there is a world of difference, and we should consider the consequences of an oxygen system failure carefully before climbing above 18,000 feet in a turbocharged but unpressurized airplane.

If going above 18,000 feet, the oxygen flow indicator should certainly be positioned where it's in the normal scan about the cockpit. If there is another person in the airplane, that person should monitor flow constantly. If two systems are available and two people are on board, they should breathe from separate systems. Having dual oxygen systems is simple enough. One can be built-in and the other portable—both available for use. It would mean only a few extra bucks and could be the most important redundancy in a high-flying airplane.

Pressurization

The ideal solution to all this is pressurization, and it's going to become more common for smaller airplanes. Seal the cabin and pump in more air than you let out, thus increasing the pressure to the same value found in normal conditions at a lower altitude. It's a rather simple thing that will become simpler with time.

Despite the advantages, a pilot flying a pressurized airplane still needs to consider the 18,000-foot level as a place to pause for thought. If there is a problem, and the cabin depressurizes at 18,000 feet, there's no overwhelming need to do something in a split second. There's time to call the controller and avoid conflicts with other airplanes. If the cabin depressurizes at 24,000 feet, the emergency oxygen masks had better work.

In air carrier operations this is recognized by an FAA requirement for quick-donning oxygen masks for crew-members when flight operations are conducted at high altitudes. These have to be immediately available and capable of being put on with one hand in five seconds.

There is a big advantage to pressurization when it comes to detecting system problems that call for a descent. Where the cessation of flow in an oxygen system would probably go unnoticed unless you are looking at the flow indicator, you'll be the first one to know about cabin depressurization. The noise, the ear sensation, and the condensation that can occur as the cabin pressure drops rapidly carry a clear message.

A primary high-altitude consideration is in taking care of the pilot and passengers. Two altitudes might be considered as benchmarks: 10,000 feet and 18,000 feet. Above 10,000 feet in an unpressurized airplane, it's best to have and use oxygen even though the regulations don't require it until 12,500 feet. Above 18,000 feet, any oxygen or pressurization system failure becomes time critical. I'd rather not fly much above 18,000 using oxygen unless a lot of extra speed or efficiency, or a lot smoother ride, results from higher flight. Even then I'd think thrice about it.

The Turbocharged Engine

The turbocharged engine is different, and one thing in particular might be bothersome on first exposure to it. When climbing and when operating at high altitudes, some turbocharged engines run hot. Most of us are used to having all the temperature needles comfortably in the center of the green sector on the gauge, and they aren't likely to be there in many turbocharged airplanes. The readings get close to the top of the green, especially when climbing or when operating at high cruise power in above-standard temperatures. The airframe and engine manufacturers are reassuring on this—they say that all's well as long as the indication is anywhere in the green. It takes a while to get used to it, though.

One reason engines run at higher temperatures up high is that there is a less dense flow of cooling air than at lower altitudes. Even though the air is colder, there's just not as much of it. Also, the induction air is heated by compression as it is rammed into the intake system to maintain sea level pressure. When you increase the temperature of the induction air, you also increase the operating temperatures of the engine. The higher you fly, the harder the turbocharger works and the greater the heat effect. Thus temperatures bear close watching. Cowl flaps (if installed) and richer-than-normal mixtures can be used to control temperatures if they start flirting with the red line.

High operating temperatures can be a factor in exposing the engine to the possibility of damage from sudden cooling. It is not good to abruptly pull the power off an engine and quickly run the temperature from hot to cold; the hotter the engine and the colder the outside air the worse it becomes.

The worst case is probably the descent in turbulence, where the objective is to keep the descent rate high and the speed relatively low. That means either using very little power or adding drag. It's sure not fuel-efficient to extend the landing gear for a descent, but if this means the difference between descending with enough power to keep the engine warm or shock-cooling the engine, the gear should be used in the descent.

Simple fact: You have to be more conscious of engine temperature and operating limits with turbocharging.

Turbocharged Flight Planning

The Pressurized Centurion (more often called the P210) that I leased has in its book a couple of figures that tell a lot about flight planning in the airplane. Maximum range at 20,000 feet is given as 910 nautical miles; it is 925 nautical miles at 10,000 feet. In other words, even though turbocharging is said to increase efficiency, this airplane will fly farther (thus get better gas mileage) at 10,000 feet than at 20,000 feet. That, of course, assumes a no-wind condition, an extremely unlikely circumstance at both those altitudes. The numbers are meaningful, however, in suggesting that the selection of an altitude must be made carefully. It is not always best to go to the altitude where the airplane will cruise fastest, because the time and fuel required to climb can nullify any true airspeed advantage at the higher altitude. Unless the airplane climbs very well or unless there's a wind advantage, it often does not pay to go too high. For what it's worth, my rule of thumb on turbocharged singles that cruise-climb at 500 to 700 feet per minute is not to exceed 10 minutes of climb for each hour of flight involved in the trip.

The case against climbing too high can be seen in another way. I hadn't had my P210 very long when I started realizing that the ground-to-ground average speed on high-altitude trips was far short of the true airspeed at cruise. Where my previous airplane, a Cardinal RG, turned in a consistent ground-to-ground average that was within 10 knots of its cruising true airspeed, the P210 was more like 20 or 25 knots shy of its cruising value. For example, on a fast 700-nautical-mile leg the P210 groundspeed varied from 220 to as high as 246 knots, yet the ground-to-ground average speed was 201 knots. Why? Almost 25 minutes was spent climbing at an indicated speed about 30 to 35 knots below the indicated speed at cruise. Then the groundspeed on the letdown diminished as I descended because of a reduced tailwind component and decreasing true airspeed at lower altitude. The increase in indicated airspeed in the descent wasn't enough to make up for this. Finally, the vector for an ILS approach (necessary because of weather) took some time. Maneuvering during departure and arrival adds as many minutes to the trip in a faster airplane as in a slower airplane, so it has a greater impact on the speed production of the faster airplane. If you turn the faster airplane around and fly it into the wind at a lower altitude, then the ground-to-ground average will be closer to its lower en route groundspeed.

Even airline jets, which climb well enough not to spend a lot of time ascending, lose about 100 knots of their cruising speed when ground-to-ground average speeds are calculated. The 727, for example, doesn't average much over 350 knots ground-to-ground in airline operation.

Wind Aloft

Keeping close tab on the winds aloft is critical to getting the most out of a turbocharged airplane. Winds don't always increase with altitude, as the forecasts often suggest, and you can find some unusual relationships between low-level and mid-level winds, especially in the summertime.

I saw a good example of this after leaving Jackson, Mississippi, in my P210 one summer day. The trip was to the northeast, and the winds were forecast to be from the southwest at 25 knots at all levels up to 18,000 feet. The weather was pretty bad over the entire route because of the remnants of a summer hurricane. It was a very tropical situation, one that would clearly be the result of a southerly flow.

My initial cruising altitude had to be 7,000 feet because of a military operating area to the northeast of Jackson. That was the highest available altitude, and the 15,000-foot cruising level I had requested could come only a hundred miles up the road.

Bouncing through some cumulus and showers, and detouring around heavier showers, I yearned for the higher altitude. But at least the groundspeed was good down low. It was 185 on a true of 160 knots, reflecting accuracy in the forecast of a tailwind. I reasoned that when I climbed higher things would really be good. The true up there would be 175 to 180 and with 25 on the tail the number would get up over 200. That is nice.

I was in for a disappointment. With the P210 level at 15,000 feet, the groundspeed actually dropped a bit, back to 180. The tailwind was gone. The southerly flow was only at the lower levels. Later in the month it happened again, only worse. A strong low-level southerly flow gave way to a light northerly flow above 10,000 feet to create a headwind at the altitude I wanted to use for my northbound turbocharged trip.

However you slice it, using a turbocharged airplane with the thought that it is "faster" than a similar unturbocharged airplane might be disappointing. The speed is there, but it is available only after you use time and fuel getting to altitude. Unless the turbocharged airplane climbs very well (which twins do and singles tend not to do) this means that a substantial percentage of the trip will be spent climbing at a relatively low airspeed and high fuel flow, and the higher true airspeed at altitude won't do much to reduce the duration of the trip unless there's more tailwind up high. I think it is safe to say you probably can't increase your average annual groundspeed by selecting the turbocharged version of an airplane.

To get value from turbocharging, planning for best altitude must be thorough and realistic lest the turbocharged airplane actually wind up delivering *lower* average speeds at higher fuel consumptions than one of its lower flying and allegedly slower brethren. If you fly a turbocharged machine, spend an hour or so with the pilot's operating handbook, a calculator, and the winds-aloft forecast for an average day. Figure trips at several altitudes in every direction. Don't forget the allowances for fuel and time to climb, as well as for descent and maneuvering to land. Guidelines on when to go high and when to stay low will become clear and you'll avoid wasting fuel and time.

Long Way Down

The first trip at 18,000 feet or more in a light airplane brings a new sensation of flight to some people. Looking down, the feeling is a lot like the one you get when looking straight down off a tall building—very high from a relatively small perch. Some have a twinge of apprehension. Perhaps this comes from the fact that you can indeed look almost straight down from a light airplane, especially a high-wing light airplane, and the relative motion of the ground is pretty slow when going 150 knots at 18,000 feet. If you find yourself avoiding the straight-down view, be assured that others have had the same reaction.

Elusive Cloud Tops

I once heard Senator Barry Goldwater do an admirable job of describing the elusive carrot held out by altitude in our quest for better flying conditions. Back in the good old days, he said, he just knew that the ability to climb all the way up to 20,000 feet would solve the weather problem once and for all. That would always be on top. Came the advent of turbocharging, and 20,000 feet was found to be far from cloud-free. Then came his jet checkout. Surely this meant all flights would be on top. Not true, though. At least it isn't true until you get into jets that will routinely operate above 40,000 feet, and even they have to go around an occasional thunderstorm. There's almost always a cloud that will top the airplane.

Even though flying in middle altitudes doesn't mean always being on top, it does help. Eighteen thousand feet can often be the difference between a comfortable easy ride and a turbulent experience.

Flying my P210 from Jackson, Tennessee, to Trenton, New Jersey, one day in May, I found a good sample of the benefits of altitude. There wasn't a great lot of weather around, but there were enough buildups to prompt deviations from course. It was minor harassment by Mother Nature. The tops were to about 25,000 feet, there wasn't enough precipitation to paint on traffic control radar, and my airborne radar showed only an occasional spot of rain. Visual examination of the clouds suggested that the ride through would be quite bumpy, though, and pilots flying at lower altitudes were continually seeking weather information from the traffic controllers. Up higher the big picture was visible. Only the tallest of the buildups were above that altitude, and they stood out like castles in the sky. Looking down, I guessed the general cloud tops in the area to be at about 10,000. I've spent a lot more time bouncing along down low in similar situations than I've spent pondering them from above, and while it's no real sweat down low, the occasional blast of rain and enthusiastic turbulence can make the flight less comfortable. I really liked my turbocharger that day.

But altitude capability won't make all flying a blue-sky experience, as was found in flying through a warm front, eastbound over the Allegheny Mountains. This was again in a P210.

The weather briefing didn't offer a lot of bad things,

but basic knowledge of warm frontal properties made me suspicious of the area. I wanted to fly through up high because of a better tailwind, and despite the fact that I should have known better, I was hoping for a flight on top of all clouds at 17,000 or 19,000 feet.

I was on top for a while, but then cloud tops came up to and above the airplane and I flew into an area of general precipitation. Most of the precipitation was topping not much above my altitude, so the weather radar in the airplane would get into ground clutter when tilted low enough to see the precipitation. This made interpretation more difficult. I could see occasional areas of heavier rain with the radar and tried a couple of deviations to avoid them.

The ride was bumpy, and it soon became icy as well. It was late spring, and I knew that I could descend and get warmer air any time; I was soon forced to that option by an increasing ice buildup. I wound up at 11,000 feet, and once through the area of weather I flew on with the feeling that the trip through hadn't been a very smooth one. A couple of colleagues flew through the same area at about the same time, at 9,000 feet, and they enjoyed almost as much tailwind component as I had up high— and their ride was smoother. I'd have been better off down low.

It was one of those systems where a 41,000-foot-or-better bizjet would have been the ticket for a velvet-smooth ride; lacking that, the average light airplane IFR altitudes were the best bets.

Singles Versus Twins

In flying high, the single-engine airplane works fine with one exception. At higher altitudes, I feel more in need of systems redundancy than down low. Thoughts of the increased distance to the ground must cause this feeling. In case of a problem, it would take longer to get down, and there would be more traffic beneath to avoid. I think a lot about whether or not the battery would run the necessary avionics for the required time after an alternator failure, and after leasing the P210 I put a lot of thought into standby systems. Hopefully the enterprising airplane modifiers of the world will have come up with a good standby system by the time you read this.

Despite the fact that speed is elusive, there is a good reason for the popularity of turbocharging. There's plenty of return for the money spent. Weather is generally easier to handle, and that includes icing as well as convective activity away from frontal zones. The big tailwinds that are occasionally snagged up high give memorable rides, and the climb ability to smooth and cool air in the summertime is like a form of air-conditioning. If you fly in the mountainous west, turbocharging can often mean the difference between going and waiting. In the east, it makes less magic. In fact, it would probably make the difference between going and waiting on very few trips. But I still like it. It about doubles the altitude flexibility of a general aviation airplane, adding important new options—and options are valuable things in flying.

13

VFR
or IFR?

Deciding between VFR and IFR for a flight where the latter is not absolutely necessary is an important item when using a general aviation airplane for transportation.

Primarily in favor of VFR is simplicity. Hop in and go. No required flight plan, no procedures, and often the routing can be direct. Primarily in favor of IFR are the procedures themselves, which handle myriad things that you must figure out for yourself when VFR, as well as the assurance of separation from other IFR airplanes.

Regardless of the benefits of IFR, opponents suggest that using the system when not absolutely necessary supports the growth of bureaucracy and increases the costs of the system. That's hard to buy for one simple reason. When the weather is bad, we expect the IFR system to be there to serve us. We grumble and growl if there are delays when the weather is IFR. If we are getting real utility out of the airplanes, trips are flown on schedule and delays are bothersome. We want to make that schedule, fair weather or foul.

There is no way for the system to know in advance which days will be good and which bad, so it can't expand and contract to satisfy need based on the actual weather conditions that exist on a given day. Using IFR when it's not really required because of weather thus can't be "wasteful." Instead, it is a means of the pilot staying sharp on IFR procedures and of getting full value from a system that is already in place.

IFR also makes a flight less vulnerable to inaccurate weather forecasting. If some unforecast clouds show up, we keep right on going as planned unless those clouds happen to be icy or cumulonimbus. If flying VFR, unforecast clouds might mean changing to a less favorable altitude, detouring, or landing and canceling the flight. The wandering we do when trying to fly VFR in marginal conditions can add a lot more distance and time to flights than moderate amounts of IFR vectoring.

Another IFR advantage is the benefit of positive separation from air carriers, which almost always operate IFR. The chances of a collision are remote, but fear of a collision with an airliner is a primary reason some companies won't let their employees fly general aviation airplanes on company business. The use of IFR is a fine argument to use if your company is reluctant on this count. If you are personally concerned with the liability (to say nothing of the discomfort of an aerial encounter with a jet—for you and for others), operating IFR shows that the pilot is exercising the highest degree of care to avoid an encounter with an airliner.

Don't think, though, that flying IFR helps avoid VFR traffic when the weather is good. Traffic advisories given by controllers are on a workload-permitting basis and are nothing to depend on. In VFR conditions, it's the pilot's job to see and avoid other traffic regardless of the flight plan.

In the unhappy event something should happen on a flight, and search-and-rescue services should be required, an IFR flight plan offers a clear advantage. The average elapsed time from the last known position to location of a downed IFR airplane was 5 hours and 30 minutes in a recent quarterly period. The elapsed time for aircraft on a VFR flight plan was 34 hours and 41 minutes, and it took an average of three and a half days to find the aircraft that were not on any sort of flight plan.

To each his own, but the advantages of IFR have led me to use it for most flights except occasional daytime trips in perfect weather.

Hazy Days

An example of how IFR can make life easier and better came as I thought back to a previous flight while maneuvering to land at Charleston, West Virginia, one hazy day.

The flight being recalled was in a new and beautiful Bonanza. It originated in Dyersburg, Tennessee, and was planned as nonstop to Trenton, New Jersey. The forecast was for good weather. There were no systems on the map, and nothing was on tap other than the usual afternoon thundershowers. I set out VFR.

About 50 miles west of Charleston, West Virginia, the weather picture worsened. It was very hazy, and the top of the haze had moved above a level I would use without oxygen. Cumulus were sprouting in the haze, and I had been quite busy dodging the fat little fellows for about 50 miles when a new dimension came to the flying. I was slogging along at 5,500 feet, and dark areas started looming in the murk. Cloud bases were lower than 5,500. This kept the Bonanza in smooth air but it also meant that I had to dodge clouds. The visibility was three or four miles, and if it was dark ahead, I reasoned that I'd have about a minute or a little more to do something between the time the dark area was perceived and the time it was reached.

The reported weather at all stations was 3,000 scattered to broken with three or four miles visibility, and I thought it best to descend to a level that would be below the clouds, hot and bumpy as that level may be. The terrain was about 1,000 feet in the area, and at 3,500 feet I was below the bases. It took a lot of squinting and squirming, but I managed to go around the first dark areas that were spotted. It was nervous business. Then I busted into a rainshower without realizing that it was there. The time had come to stop the VFR foolishness. I found a reasonable place to circle and started the tedious process of air-filing an IFR flight plan and getting a clearance. It took about 30 minutes, but I felt a lot better about things as I climbed to 7,000 feet and moved on toward the destination.

The day of the second flight was a carbon copy of the previous one except for a bit more thunderstorm activity. The big ones were popping up all over, but I had airborne weather radar and it was doing a good job of outlining the flight path. Still, the flight could have been safely conducted without it. The top of the haze was well below the 19,000-foot cruising level of the turbocharged airplane, and the cumulonimbus were beautiful against the bright blue sky above the haze.

When it came time to start down, I could see that there wouldn't be any serious problems getting to the Charleston airport, though it might be bumpy. Some tops ahead were at or above 15,000 feet and appeared to be building rapidly.

Down lower, I was forcefully reminded of the VFR trip just related. All stations were reporting VFR with a surface visibility of three or four miles, and the flight visibility certainly wasn't any better in the lower levels. The cumulus were floating by, and occasionally I'd fly into one that would have been difficult to avoid even if I had tried. Some showers ahead were shown on radar as the descent continued, but it was a simple matter of deviating. The radar in the airplane was being used, but information obtained from the controller's radar would have given almost as good a ride. I say "almost" because he wasn't painting all the cells that were showing on the airplane's radar.

Other airplanes are as much concern as weather when near a terminal area in marginal visibility conditions, and at one point there was some discussion with the controller about zigging or zagging. I wanted to fly a heading of 060 because of weather, he wanted me to fly 030. That was into a cell, though, and we finally compromised on 360 to go around the north side. When he had time, the controller explained that Charleston departures were going around the south side of the weather, and we'd have been flying right at them had we done the same. The flying was largely by reference to instruments even though the meteorological conditions were legally visual most of the time, and looking for other airplanes in that murk was a tough chore. I could pick a path that would avoid the weather; I was pleased to have some help from someone on a path that would avoid both the weather and jets climbing in the opposite direction. The closure rate with a climbing airliner would have been about 670 feet per second, and that would make the see-and-be-seen concept difficult in the marginal conditions that existed.

On the VFR trip through this area, I had been wandering aimlessly, avoiding clouds. Even though I wasn't landing at Charleston, I would have passed close by. Had I exercised the option to remain silent (and clear of their airport traffic area), it would have been perfectly legal. But it wouldn't have been such a good idea, as illustrated by the IFR flight. There was a very valid reason—a climbing airplane full of people like you and me—not to fly down the path I would have normally chosen, and the use of the IFR system provided separation from that airplane.

The System Works

One rainy Saturday I found a good example of how the IFR system works even when the weather is bad in a major metropolitan area. The mission was to go from Trenton to Teterboro, New Jersey, and make a local flight from Teterboro in a Merlin IIIB turboprop that I was evaluating. I rented an A36 Bonanza for the short trip.

Teterboro is under a layer of the New York TCA, and IFR traffic there is controlled by the "Common I," a terminal radar control facility that handles all the New York airports. Teterboro itself is one of the busier airports in the nation, and I thought I might be incredibly optimistic when I called for an IFR clearance only 30 minutes before an appointment at Teterboro, 45 miles away.

The clearance came through instantly, though, and the Bonanza was soon at 3,000 feet, moving through the clouds and showers. The vector was for a straight-in ILS approach, without delay. The Merlin flight was handled out and back into Teterboro quite efficiently, and my Bonanza ride back to Trenton was the same story. There was a hint that VFR would have been possible for the Bonanza on this trip. Conditions were alternating above and below 1,000 and three, and the forecasts called for better conditions. The forecasts were wrong all day, but viewed through rose-colored glasses they might have suggested VFR. It might have worked with a Special VFR at Trenton and Teterboro (in the Bonanza—no way to do that in the Merlin), but why accept the risk of scud-running? Even if there had been a 30- or 40-minute delay for IFR, it would have been a better deal.

The advantages of flying IFR to a busy airport were detailed in Chapter 9, and some advantages of using IFR charts and procedures at night were covered in Chapter 10. The advantages can extend to any day or night flight. In fact, although IFR is thought of as being difficult, it is often easier than VFR. Follow a cross-country flight through from start to finish and consider some IFR advantages.

Planning

Planning an IFR flight is a matter of determining which airways go most directly from here to there. If area navigation gear is aboard, all that's necessary is to draw a straight line on the chart and pick off two or three way points. Planning a VFR flight can be much the same, but when it comes to matching the route with the weather, there's a big difference.

When checking weather for an IFR flight, the primary interest is in the synopsis plus actual and forecast weather for the departure, destination, and alternate airports. Add to this the need to know about any en route thunderstorms (which should be shown on the radar summary chart) or icing conditions and you have a framework of the necessities. For a VFR flight, you need in addition some idea about clouds and obstructions to visibility en route. The clouds that are flown through routinely on an IFR flight must be avoided on a VFR flight. And, unhappily, information on en route weather (except at reporting stations) is not available except in the form of radar reports, which show rain, not clouds.

Most IFR flights are cleared as filed. Flying VFR as filed (or desired if a flight plan isn't filed) depends on the weather. You can go as you wish only if the weather is as you thought it would be.

An IFR flight might involve a departure procedure that would seem wasteful to the confirmed VFR pilot. But the procedure is usually for the purpose of avoiding obstructions—a worthwhile goal whether VFR or IFR.

As we get into the IFR flight, we are doing something that has organization to it. The task is there for study in the form of the clearance, and sparse as they may appear to some, the IFR charts contain a lot of information. Distances and minimum safe altitudes for the airways are all there. Approach plates are nothing more than a set of detailed directions on how to approach an airport safely and position the airplane for landing. (Visual charts should be carried on all flights, though, in case the need arises to proceed visually.)

When an IFR flight is in cloud, there is some additional workload to flying by reference to instruments, or monitoring the autopilot, but this is offset to some extent by removal of the need to watch for other aircraft when actually in cloud.

The ease of navigation can be equal VFR or IFR, but only when VFR is flown at reasonable altitudes. An IFR flight is always at an altitude where the appropriate navigational aids can be received; if the VFR has to be down low because of clouds, that's not true.

The arrival is a prime time of IFR advantage, especially if the weather is the least bad or the airport is a hard one to find. The instrument approach is a matter of matching the numbers on the instrument panel with the numbers on the charts.

Freedom of Choice

All this unrestrained enthusiasm for IFR must be taken with grains of salt, for there are times when weather conditions actually favor VFR, but VFR flying in these cases is likely to be very demanding, as I found one day in the Middle West.

There was some icing on the approach to Quincy,

Illinois, so I was suspicious of IFR for the following leg, to the east. The weather at Quincy was VFR, and it was forecast to be so to the east. Barely so, but still better than 1,000 and three. I launched VFR, to stay low and out of the ice. The ceiling along the way was lower than 1,000 feet at times, but the visibility was okay. The best route was to follow a highway, but there wasn't one that defined a good path. Some improvising was necessary.

Two TV towers demanded attention. One, west of Jacksonville, Illinois, wasn't bad. The ceiling and visibility were at their best in that particular area, and the tower was visible from a position over a river to its west. The next one was tougher. The ceiling was low in the vicinity of Springfield. I wanted to give the airport a good berth, so a flight path around the southern edge of a lake to the south of town was chosen. From there I wanted to catch the highway that runs from Springfield to Decatur, but a 1,458-foot TV tower stood in the way. I used a radial of the VOR as a backstop—no crossing the radial unless the tower was in sight—and just before reaching the point I saw the tower and its guy wires. The top half was invisible in cloud, and even though I was a couple of miles away, I felt as if the guy wires were reaching out for my airplane. Maybe IFR with a little ice would have been better. Or how about a nice motel back in Quincy?

I kept pushing on VFR, but the flight terminated of necessity at Indianapolis. A freezing rain had commenced. In retrospect, it was a lousy flight. Very uncomfortable.

A better use of VFR in preference to IFR came while negotiating with a front between Kansas City and Wichita. There were some thunderstorms, and the controller's radar was doing a good job of picking them out. Still, it was uncomfortable at 10,000, then 8,000, then 6,000. There was a good collection of building cumulus independent of the storms, and these were working the airplane over with a vengeance. The Wichita weather was good, though, as was the weather behind, so when a broad opening in the clouds appeared, I canceled IFR and descended VFR to a position below the clouds. There I could eyeball the rain shafts, and the ride was much smoother. In fact, the only bumps at low level came right in the frontal zone, where the wind was shifting from southerly to northwesterly.

Canceling IFR and going VFR isn't without hazards. During a controller's strike in another country, a pilot canceled IFR and proceeded VFR after being told to expect a long delay. He hit a television tower on the VFR leg of his flight. That emphasizes the necessity of having visual charts along. Recognize, too, that the transition from instrument to visual flight is demanding. When switching to VFR, we have to know present position, we have to have visual charts and study the route to be followed, and we must interpret the weather as it applies to VFR flying. No question, it's better to plan and fly a flight either VFR or IFR from start to finish, but there are times when we'll want to improvise on this and switch from one to the other. Just be aware of the requirements when changing.

14

Those Magnificent Machines

All this talk about flying depends on having an airplane to fly, and the choices are many. There are far more basic shapes of airplanes than cars, and the reasons pilots prefer what they own, lease, or rent are as numerous as pilots themselves.

An airplane does not have to be big and fancy to deliver excellent transportation. Nor does it have to be expensive to provide good transportation at minimum risk. Pilot skill and judgment weigh heavily here, and flying can be both as productive and as safe as possible on a limited (for aviation) budget—if the pilot is both proficient and careful.

The Author's Selections

Over the years I've felt close to my airplanes. Most have been true friends and have served faithfully. There have been adversary relationships, too, but they usually don't last long. One thing I learned quickly was that if the use of an airplane is concentrated in a type of flying that best suits the airplane, things go more smoothly.

Piper Super Cruiser

The first airplane that I bought was a Piper Super Cruiser. It was a '47 model; I bought it in 1952. The airplane had been used on pipeline patrol, so it had a lot of hours, but it seemed to be in good shape. I paid $1,200; a Bonanza of the same vintage was available at the time for about $5,000. Rare bargains.

The first uses of my Cruiser were localized. I took my flight instructor checkride in the airplane, and used it mostly in running around the state of Arkansas, where I was living and working at the time. There were a couple of 1,400-foot strips that I used occasionally—one for fishing and the other for visiting kin—and to me the airplane was just ideal. The 115-horsepower engine, the old Cub wing, and the big fat tires made it a natural. I was a happy boy.

The next event was a move to Moultrie, Georgia, to work at an Air Force contract flight training school. I went there in my Cruiser. This was at a time when hangar rent was relatively inexpensive. I had never left a fabric-

covered airplane out of doors, and I didn't want to start.
The closest hangar I could find was some miles away, in
Tifton, Georgia. That was inconvenient, and it planted
the seed of desire for a metal airplane that I could tie out
at Moultrie. My situation was no longer ideally suited to
the Cruiser, and this just naturally prompted thought of
something else even though the Cruiser was doing well
in the primary mission.

There were about 460 nautical miles between where
I worked and where I wanted to be every weekend, and
my airplane's work was to cover that distance as often as
possible. For an airplane that cruised at about 100 (stat-
ute) miles per hour, the Super Cruiser was amazingly
adaptable to this need. The 36 gallons of fuel and the
cruise consumption of five and a half gallons per hour
made possible nonstop trips when the wind was favorable.

This was before the time of IFR avionics for light
airplanes, so flying was strictly VFR. The Super Cruiser
was a pretty good scud-runner, too. The relatively slow
speed and good visibility over the nose helped, and for
time of need I put a turn and bank in the panel. It got
used a couple of times, when I pushed too hard.

Swift

Like anyone using an airplane to travel, I found the
lure of more speed to be strong; this lure, combined with
my desire for an airplane I could leave outside,
prompted forays into the market. I looked at a Bellanca
but decided that wood and fabric wouldn't live outside

any better than all fabric. Then a shiny and sleek all-metal Swift caught my eye. I met the owner of the Swift and discovered that his airplane at least made the pretense of fitting my missions better than the Super Cruiser, and the reverse was true as well. He needed short-field ability. We flew each other's airplanes, and in the course of doing this I told him that the Super Cruiser would take off in only a few hundred feet. He asked if it would fly away in the distance across the ramp. I replied in the affirmative, and a few minutes later found myself proving it to him. He was impressed, and on my paying him a little to boot, the trade was made.

Once I had the Swift, I began to learn that glitter doesn't necessarily relate to gold. The Swift held only 25 gallons of fuel and it burned about seven gallons per hour. In three hours the gauge was in a bad place. The airspeed indication tended to verify the high cruising speed that had been quoted, but I could never seem to get much more than 100 to 105 knots out of the airplane. A nonstop over my 460-nm run was impossible. With a strong headwind, a one-stop was touchy at best. And where the Swift was definitely faster than the Super Cruiser, the elapsed time in it was often greater because of the requirement to stop. The Swift would do in four and a half to five hours flying time what the Cruiser had been doing in five to five and a half, and the average 30-minute advantage was more than shot in at least one fuel stop.

I had learned a lesson about the relationship between speed and endurance: The former can be seriously handicapped by a lack of the latter. I had also learned something about carrying capacity: The trips with another person along strained the Swift's lifting ability, whereas the three-place Cruiser had done very well with a passenger. It was my first brush with the fact that it's best to have an airplane with one or two more seats than you normally use. I was glad to get rid of the Swift. Appearance was its primary asset, and I found this of little value in traveling.

IFR-Equipped Piper Pacer

After the Swift came my first real airplane, an IFR-equipped Piper Pacer. Bought from my father in 1955 at a nepotistic price, the Pacer was to serve well as I got my instrument rating, as I learned to use that rating, as I flew my way through an aerial courtship, and as I started off in the aviation magazine business.

The things that people seek in an IFR airplane were the same then as now. There was a single-axis autopilot, a rarity in the mid-fifties, three separate radios (a VOR, an ADF, and a low-frequency receiver—all three had a transmitter), and a battery-operated standby radio on the floor. A full gyro panel completed the array, and there was even an alternate vacuum source for the turn and bank.

If the Pacer was short anything, it was gasoline. The tanks held 36 gallons (the same as the Super Cruiser), and the consumption at 100 knots was just under six gallons per hour. That wasn't bad with no headwind, but going into the breeze it didn't transform into a lot of miles. My father had tried to help this problem. There was a fitting under the left wing, and he had a five-gallon can with a hose, a matching fitting, and a pump to use in adding five gallons to the left tank in flight. I think he used it only once. I never used it; the rig struck me as being somewhat questionable. Instead, I just made some extra fuel stops.

The Pacer taught me the value of IFR equipment and the ability to use it. I finally had a little airplane that offered some schedule reliability without engaging in the very risky business of scud-running. By this time I was in the Army, stationed at Fort Rucker, Alabama (which took about 80 miles off my familiar run back to Arkansas), and the schedule reliability was important. Arkansas was far outside the allowable limits for trooper travel on a three-day pass, and if I went there on a long weekend, I surely had to get back on time. It wasn't like business flying, where a big deal might have hinged on being there. Instead, the airplane getting me back on schedule one more time meant depriving the company commander of the pleasure of cutting the stripes from my sleeve and throwing them in the trash. To say nothing of all those potatoes I'd have to peel. The log entry for a lot of the flights included actual instrument time. One showed three and a half of the six hours spent en route as being flown in cloud. I enjoyed the challenge.

The Pacer also had a good short- and soft-field capability if flown at reasonable weights. To go sunning and swimming, we used a little sandy strip across the street from the beach at Panama City, and the Pacer flew well there with a modest fuel load and as many as three on board. A good on-shore breeze and the big fat Cub tires helped. To have an airplane that was as at home IFR as on a sandy strip was, to me, a fine thing. That a lot of people thought I was crazy for flying IFR in a light airplane was of no bother at all. They just couldn't fathom

being inside a cloud in that collection of bedsheet and
tubing; I liked it and wasn't about to upset their vision
by telling them that the Pacer was nice and cozy when
inside cloud.

After I got out of the Army, my relationship with the
Pacer was put to an interesting test. I was going to col-
lege and also had a job flying a fine Beech Twin Bonanza.
Additionally there was a romantic attachment back close
to Fort Rucker. That was a switch. After spending several
years in Alabama and Georgia running back to Arkansas,
I now found myself in Arkansas running back to Alabama. That's what airplanes are for.

A lot of my friends asked how I could stand to flog
my Pacer through dark and rainy skies after moving
smoothly and rapidly about in the big Beech. It was easy.
The Pacer went where I wanted it to go. It was mine, and
I enjoyed it. The Beech went where my boss, a fine man
by the name of Ben Hogan, wanted it to go. It was his
and he enjoyed it.

For some reason, the weather seemed to worsen in
this period of courtship. The trips to Alabama were IFR
all the way. And I wasted one trip. The tale has been told
before, but it's worth repeating because it's reflective of
the feeling and friendship that were found in the small
aviation fraternity in the fifties.

It was a dark and stormy night, and I was IFR headed
for Dothan, Alabama, in my Pacer. I had bought a ring
the day before and was taking it to my true love. This was
before the days of direct pilot-to-controller radio contact. The drill then was to report over each point—VOR
or low-frequency station—to what is now the flight service station. I had flown the route enough to have radio
pals along the way and had one in the station at Green-

wood, Mississippi. The conversation this night went about as follows:

N125 RC: Greenwood radio, 125 RC is over Greenwood at five zero, level at 5,000, estimating Meridian at five five after the next hour, Evergreen after that.

Greenwood: Roger 125 RC and, ah, Richard, what are you doing out on a bad night like this?

N125 RC: Well, I'm going over to propose to my true love.

Greenwood: Oh, don't you want to talk about that? I'll get you clearance for an approach if you want to land and talk about it.

I didn't land, but neither did I get the ring out of my pocket that weekend. It took another trip, some days later, to make the deal.

The Pacer got an interesting IFR workout in getting to our wedding, too. There was a record snowstorm in the southeast, and I had to fly through it to get there. It seemed ludicrous to find that much snow that far south, but then anything can happen to a young man on the day before his marriage.

The Pacer hooked me on instrument flying. I had seen what it takes to make an airplane serve as reliable transportation, and it was with a heavy heart that I sold the airplane to raise money for a down payment on a first house.

To this day, I still compare airplanes with the Pacer. It's interesting to match the time en route in the finest new machines against that faithful old bird, and while the new ones are almost always faster and more comfortable, the Pacer did pretty well.

250 Comanche

The next airplane I flew a lot after the Pacer was one of the first 250 Comanches. It was a good airplane, but it had the Swift's problem with limited fuel capacity. With 60 gallons it was good for only about three hours if cruise was at 75 percent power and if every landing was to be with at least an hour in the tank. On one of the first trips in the airplane I remember being quite disappointed with the short distance covered before the gauges were down in an uncomfortably low area. At low power it would fly longer, of course, but I seemed intent on going faster instead of slower. That gave way to some conservatism on power, and I started routinely getting four-hour hops out of the Comanche with an hour's reserve.

The first "long-range" airplane I used was also a 250 Comanche. With the '61, Piper increased the 250 Comanche's fuel to 90 gallons, in two 30- and two 15-gallon tanks. It was, to me, the best arrangement for fuel as far as maximum-range flying goes. Landings were approved on any tank, so I'd keep one of the 15-gallon tanks in reserve—it would be good for more than one hour—and have an absolute determination of fuel reserve. The engine on the airplane was carbureted, and running a tank dry without losing the engine completely was easy if you'd pay close attention to the fuel pressure gauge and switch tanks as soon as it started flickering. I routinely flew this airplane on six- or six-and-a-half-hour hops (at 60 percent power) with over an hour's fuel in reserve. This would cover up to 1,000 nautical miles with a little tailwind. It was my first exposure to an airplane with this much range (the Twin Bonanza I flew didn't have such long legs), and the feature came on strong as a good one for any airplane.

Twin Comanche

After a few years of single-engine Comanches, the Twin Comanche came along. This was a first experience with a twin to take me where *I* wanted to go, and I did the predictable things with the airplane. I tried flying more at night but found that the limiting factor in my night flying was fatigue, not the number of engines on an airplane. After working all day, I simply didn't feel like flying all night. Being an early riser probably contributed to this. I made a few night trips in the twin and then went back to the usual pattern. I also probably tried to fly more "weather" with the twin right at first, but I quickly learned that I had been flying an entirely adequate amount of weather in the single and that the weather didn't care how many engines were on my airplane. It was at this time that I developed the strong feeling that a pilot is looking to get his nose bloodied when he starts doing things in a light twin that he wouldn't do in a single.

One of the Twin Comanche's features was its ability to use small fields. It had to be flown like a single when doing this—if one engine failed there would be no alternative but to shut the other engine down until the airplane was safely out of the field and a few hundred feet high. Accepting that risk, I used the Twin Comanche regularly from one of the short strips that I used to visit in my Super Cruiser. I always operated the twin at very light weight, and the flying was without sweat or incident. Looking back, I feel it was unwise—if I had a twin today I would always try to fly it with engine-out safety margins.

One of the first turbocharged Twin Comanches was around for a while, and with it I sampled some of the benefits of high altitude flying. It was like night flying, though—something done only a few times. Having to wear oxygen was not desirable, and the engines ran hot up high, requiring extra fuel for cooling. This torpedoed the range of the airplane even though it had tip tanks and carried a total of 120 gallons. To me, it was a hybrid airplane—a single-engine Comanche converted to a twin, converted to a turbocharged, converted to a long-range airplane with the addition of tip tanks. It had a lot of flexibility and performance, but I was never comfortable with it.

Skylane and Cherokee Six

I had a Skylane for a while, which is as reliable as a good horse. It didn't have the range I would have liked, even with long-range tanks, and was a little restricted on weight for the load that I wanted to carry. I traded it for a Cherokee Six, which did a masterful job of load carrying, but after a couple of years I found myself flying in it alone most of the time and sold it.

Cessna Skyhawk

Next came what I thought was an interesting experiment. I bought a new Cessna Skyhawk plus a very complete array of King avionics, including area navigation gear and a horizontal situation indicator in addition to the regular directional gyro. It was to evolve into what I considered an ideal instrument panel, and the best arrangement I have ever seen in a light airplane. There was real redundancy, with the electrically operated HSI and the vacuum directional gyro giving reliable heading information after either a vacuum or electrical failure. I could fly with a pretty full deck after a failure of either power source. The Skyhawk was great fun to operate IFR, too, with instrument approaches to minimums very nice to fly, what with the HSI and the Hawk's good handling qualities.

It was slow—a cruise of 120 knots was about as good as it would get—but it was my modern-day version of the Pacer: a basic, simple, and reliable airplane with as elaborate an avionics package as possible. Since my average trip at that time took me 300 nautical miles from home, the relatively slow speed of the Hawk didn't hurt too much. A lot of the flying was done in the windy Great Plains, and the airplane behaved remarkably well on the ground in the wind.

When I bought the Hawk I wondered if I'd be able to complete trips on as good a schedule as I had been doing in the Skylane and Cherokee Six, and the answer was yes. More slowly, to be sure, but it was as good for IFR operation as the others. I flew it more at night than I had flown in a long time, too, and found that I even felt comfortable with it when contemplating the unlikely engine failure. The Hawk's landing speed was very low and I thought that the consequences of any night forced landing would more likely than not be bearable—especially considering the use of shoulder harnesses.

There were some engine problems with the Hawk as it passed 1,000 hours—largely due to the use of 100-octane fuel in an engine designed for 80—and I reverted to the natural pastime of thinking about a somewhat faster airplane. One thing I quickly learned is that the best way to get the money out of an overequipped airplane is to fly it out. When testing the market with the Hawk, I found that it would bring only a thousand or two more dollars than a Hawk with the standard array of a couple of nav/com radios and an ADF. Just like a house, an airplane that has been made the fanciest one around must be allowed to give return in the form of use rather than in the form of increased selling value. I'd add that the return in the form of use is usually well worth the cost.

Cardinal RG

My next airplane was a Cardinal RG. It was very efficient, and in the 1,200 hours that I flew it I came to have a great respect for this retractable-gear airplane with a 200-horsepower engine. It yields a high percentage of the travel capability of any light airplane, including light twins, at a much smaller percentage of the cost.

Centurion P210

After the Cardinal came the leased Pressurized Centurion, or P210. Having been an advocate of a full-capability single, I couldn't resist the opportunity to use one of these. The airplane has radar and is equipped for flight into known icing conditions, and the pressurization is very nice. First impressions are never as good as those formed after a couple of years of flying, but in the first 200 hours I felt the strongest suit of the P210 to be comfort. It is very quiet and smooth. All-day trips are not fatiguing. I soon learned not to waste a lot of fuel and time trying to vault up to a high altitude on every trip. The pressurization can serve as well on lower-altitude trips as on higher: When the airplane is cruising at 8,000 feet, the cabin can be down at 800 feet, making the cabin of the airplane the same as the environment at home.

The P210 is like a miniature airliner, but despite its equipment and comfort, it is still a light airplane. I had flown it about 200 hours when a line of weather offered a lesson on this.

I was cruising at 15,000 feet and could see much higher tops ahead. The radar was showing some fairly heavy stuff, but it didn't seem to take any well-defined form. Closer to the rain area I got a clearance to descend to 13,000 feet to be comfortably below the freezing level, and as the rain started I asked for further descent to 11,000 feet.

The controller said it was about 50 miles through and that some of the return appeared heavy on his scope. He did add that a 727 had just gone through with no problem.

Once I was in the rain area, attenuation (the dissipation of radar effectiveness caused by being in rain) took a toll, and I felt I was getting only a five- or ten-mile range out of the radar. And there were contour areas close ahead, indicating very heavy rainfall. To me the best heading appeared to be about 110 degrees. The controller called and said that to him the best heading looked like 75 degrees. That looked terrible on my scope. Maybe the 727 did go through okay, but this was no 727: I made the decision to retreat. There was simply too much conflict in the available information to feel comfortable in continuing.

I learned from the Twin Comanche that weather doesn't care about the number of engines on an airplane. This experience was one step in learning how weather responds to the fine array of equipment on the P210. Radar, deicing, and altitude capability are truly wonderful things, but each must be used with care.

This summary of airplanes used by one person for transportation is offered only as an example of motivations in airplane selection. Some brands of airplanes are not represented, but that is only because the proposed mission and the available money didn't fit at the time. Two I hope to eventually have in my hangar are the Bonanza A36 and the Mooney 201. By the time I get to them they might be in more exotic forms than at present, which is all to the good; like any pilot, I can only select the airplane that best fits the pocketbook and the need of the moment.

Used Airplanes

Used airplanes can be good bargains, but they must be bought with full understanding of possible additional expenses.

First, the upkeep is keyed to the price of a similar machine bought new. After setting aside the purchase price, it will cost as much to fly a 10-year-old Bonanza as to fly a new one—and it may cost more. If you select an old and low-priced twin, the operating and maintenance costs will be the same as for a new twin. A lot of people have found this an expensive lesson to learn.

Second, components on a used airplane may not go to their life expectancy. Even though an engine has 700 hours to go to the published time between overhauls, it might not make the grade for one reason or another. Nor might the vacuum pump or alternator or instruments or other items.

All this means that the money you save by buying used might not stay in the bank. The buyer must be ready to cover expenses that are by nature a risk of buying a used airplane.

The Final Selection

Many final selections of an airplane are based on emotional considerations, such as ego satisfaction and aesthetics (in the eye of the beholder). We owe it to ourselves to make a thorough analysis of an airplane before choosing it, to make sure the satisfyingly beautiful machine will meet the goals set for it. Beauty can fade fast if the airplane won't carry the load, go the distance, fly from the airport, top the highest hill along the way, or operate on a reasonable amount of fuel. If those things are satisfied, the relationship with an airplane can be beautiful.

15
Fine
Points

A pilot's ability to fly the airplane is not directly related to flying time, and the *new* pilot seeking knowledge based on 1,000 hours' experience seeks an elusive goal unless he makes a proper effort. It is not automatic: Some 300-hour pilots are airplane-wise beyond their experience; some 3,000-hour pilots are retarded in understanding the fine points of flying.

Definition

Fine points might be defined as the things that reduce risk and make flying easier while at the same time enhancing the utility of the airplane. Some are procedural, some personal. All require thought and an open mind. The list of fine points should be endless, to be continually expanded by each aviator. I've covered a lot in this and other books, but there are plenty left. The key is in studying each question and event, to take advantage of every possible lesson.

Risk Reduction

In an article in *Flying* about making decisions on weather I mentioned that under some circumstances I would make a very low visibility takeoff. The conditions: reasonable terrain, an uncongested area around the airport, and no nonpilot passengers. At least one reader wrote in saying that I had taken leave of my senses. To him, a low-visibility (or zero-zero) takeoff would be unthinkable.

The letter led me to reexamine the fine points of risk in this situation. In making a low-visibility takeoff at Wichita, Kansas, the advantage of flat terrain is strong. If the engine fails on a single, or if an engine fails on a twin, a reasonable wings-level descent at a proper airspeed would minimize the risk involved. Sure, you could be unlucky and fly into a grain silo broadside. You could also settle gently into a wheat field. Is the risk there any

greater than on a clear-day takeoff at an airport in a
congested metropolitan area?

I used to fly from Linden Airport in New Jersey, a
field that is surrounded by factories, oil tanks, refineries,
houses, a cemetery, and other items that would be un-
friendly to a powerless aviator. Certainly the options in
case of an engine failure on takeoff at that airport on a
clear day would be less than after a south takeoff at
Wichita's municipal airport with a 100-foot ceiling and a
quarter-mile visibility.

Interestingly, forced landings at Linden have not
been a major problem. Engines must remain faithful to
their pilots. Or perhaps the pilots, recognizing the un-
friendly area around the airport, do an extra good job of
draining sumps, checking tanks, and having the airplanes
maintained. Likewise, there is no history of trouble with
low-visibility takeoffs by experienced pilots flying well-
maintained airplanes. But a pilot should carefully examine the risks in either situation. And if flying from the
congested area airport on a day with very low conditions,
a wise pilot might decide that the two risks together add
up to a larger risk than is acceptable.

This was fresh in mind when I stopped at Wheeling,
West Virginia, in my P210. That airport is on a high spot,
with deep gullies and low hills around it—a bad place for
an off-airport landing. The day was hot, the load was
heavy, and the fully loaded P210 is not noted for its
jackrabbit acceleration and rocketlike climb on hot days.
Lifting off at 70 knots, the recommended speed, resulted
in a little bleep from the stall horn, and the airplane
wasn't going up too well as the last of the runway passed
beneath. Not a lot ahead, and for interest I started picking the places I would deposit the airplane if a power

failure occurred. None was good, but any would have accommodated an arrival, hopefully with no more than some bending of the airframe. Later, at the end of the leg, I made a normal approach, landed at home base, and turned off at an intersection 1,150 feet down the runway. Even though the airplane seems "hot" on takeoff, it will get down in a short space. That would help minimize the risk involved in an off-airport landing.

There's no way to make the risk on every takeoff equal unless you always fly the same airplane from the same airport in the same kind of weather. The fine point is to evaluate each risk carefully, and in advance.

Alertness

Staying alert when flying is another fine point. If lethargy is allowed to sweep over the brain, risk-evaluating suffers—and so does flying. It takes a special effort to be as alert at the end of a long flight as at the beginning, but it's worth the effort.

More insidious is a complacent state that leads to a relaxing of the standards of flying. A study has revealed that in a surprisingly high number of accidents at least one crewmember was whistling as events unfolded. What does whistling mean? People might whistle when they are apprehensive, but it's more likely that they whistle when relaxed. I like to whistle at pleasant times. Strolling on a nice evening is a good time. When I've been tempted to whistle while aloft, it has more often been when flying a strange airplane or flying in unfamiliar circumstances. The urge to whistle comes when I think

I've got it pretty well figured out. Phooey with that. Knowledge of the prevalence of cockpit whistling before accidents has led me to become very suspicious of what's going on any time I start to whistle in an airplane. If tempted to whistle while flying, consider that you can't whistle and think at the same time.

Safeguards Against Distraction

Guarding against distraction is another fine point. An air carrier collision with a general aviation airplane offers food for thought on this subject.

There was an extra crewmember in the cockpit of the airliner, and this person was carrying on a conversation with the captain as the airplane maneuvered for landing. They talked about lawyers and money and about mistakes made on a previous flight. There was laughter; apparently a good time was being had on the flight deck —all intermingled with the prelanding checklist. At the exact time the other airplane (the one the airliner eventually collided with) was first mentioned by the controller to another flight, the off-duty crewmember was chatting away. At the time it was first mentioned to the ill-fated flight, there was background laughter on the flight deck. A good story and laughter preceded the next call of the traffic by five seconds, but after this call the crew said they had the airplane in sight.

When the traffic was called again, at one mile, the flight deck discussion became serious. But then there was more laughter on the flight deck 30 seconds before the airliner hit the slower general aviation airplane from behind, and a terrible accident was carved into history.

According to a National Transportation Safety Board study, the general aviation airplane would have been visible from both the left and right seats in the airliner cockpit for 170 seconds before the collision if both pilots were seated so that their eyes were at the design eye reference point. Ten seconds before the collision it should have been clearly visible with the crewmembers seated normally, and even easier to see had they leaned forward a bit, to what the NTSB calls the alert position. (And surely pilots who were supposed to be following an airplane visually would be in the alert position when they realized they had lost sight of the airplane.)

The NTSB did not choose to include the distractions in the cockpit as a factor in the accident, but we should at least relate the subject to our flying.

Idle chit-chat with another person during flight makes it quite easy to miss a radio call or to misunderstand what is said. When another person is talking to you, chances are they won't hear a radio call to you and will drone right on through a communication from the ground, making it difficult for you to comprehend the controller.

Continuity of thought can be lost to distractions. If you are tracking a thought, such as one about another airplane, and then have a hearty laugh at a good story, it isn't easy to go back to the business of thinking about and tracking the other airplane. Where was it? Finding it a second time can be as difficult as finding it the first time.

At least one airline captain that I know of deals with this decisively. He makes it plain that the only talking in the cockpit when below 18,000 feet will be in relation to the safe conduct of the flight. Some might feel he is a tyrant, but the value of his policy has been proved.

Another airline accident involved crewmembers who were talking politics as they busted through altitudes without noticing and then flew into the ground; still another involved a crew that was preoccupied with a discussion of possible activities in a parked car off the end of the runway.

There's a time for flying and a time for chatting and joking. The latter might be okay in noncritical phases of a flight, and a fine point of flying technique is in knowing when to tell everyone to be quiet so you can concentrate on the task at hand. If bothered by chatty passengers, there's always retreat into the solitude of good headphones.

Self-Check

Another fine point is to run a continuous check ride on yourself. This might seem a friendly way to do it, but it works. If you make a lousy landing, a little thought will reveal the reason as well as any flight instructor might. The only requirement is that you know the principles involved. Was the airplane at the proper approach speed over the fence? Was the airplane in the landing attitude at touchdown? Was the final foot of altitude lost gradually? Everything you do in flying is more visible to yourself than to anyone else, because only a pilot knows where he is looking and what he is thinking. It takes a conscious effort to grade yourself, but it pays dividends for you accept nothing short of perfection without going through an analysis of why it wasn't done as well as it could or should have been done.

Finest Point of All

Switching from fine points to fine things, consider for a moment that people have never had available to them anything that offers as many good things as does the general aviation airplane. It fulfills the age-old dream of flying like a bird. Go where you want to go, when you want to go. Flying is enjoyable and challenging, and it involves an acceptance of responsibility that is unique in this modern world. This helps build character in the young and offers renewal for the old.

During the bicentennial I interviewed a couple of teenage fellows who were flying a Skylane around the periphery of the United States. We had agreed to meet in Mobile, Alabama, and as I flew there to meet them, the weather was less than cooperative. Thunderstorms were abundant, and on arrival at Mobile it was an approach to minimums. The two teenagers beat me there.

In response to my account of this, someone wrote that they thought it ridiculous for a couple of kids to be out risking their hides and a perfectly good airplane shooting approaches to minimums. I couldn't have disagreed more. It was a sign to me that they were capable people. Well trained and confident, they were able to accept responsibility and to perform when called on.

It is indeed a wonderful privilege to live in a country where young and old alike are free to learn to fly, to excel, and to use an airplane. I often sit with my hands on the controls of the airplane, moving along from one place to another, and reflect for a moment on how wonderful it is for an old country boy to be able to do this. Then, reverie broken, I get back to thoughts of flying as precisely as possible, for it is by doing it well that we protect our privileges of flight. In time we come to realize that statistical averages on airplane accidents do not represent inescapable hazards in flight. The averages are, instead, the scoreboard on the careless, the reckless, and on those who don't care enough to do their very best.

Index